CORPORATE SOLDIERS AND INTERNATIONAL SECURITY

This book is the first to explore comprehensively the rise of UK private military companies on the international stage. After illustrating early examples of private force, such as mercenary companies, which filled the ranks of European armies right up to the 1850s, Christopher Kinsey traces the development of UK private military companies (PMCs) from the mercenary organisations that operated in Africa in the 1960s and early 1970s, through to the start of the rise of legally established private military companies in the mid-1970s and early 1980s, to today's private military companies which are now important contributors to international security and post-conflict reconstruction.

This volume first points to why and how the change from the mercenary organisations of the 1960s and 1970s came about. In particular it examines in some detail the Sandline affair, a crucial turning point in the industry's history. It then analyses how PMCs have been able to impact upon international security. Finally, the book examines the type of problems, as well as advantages, that can arise for organisations that decide to turn to private military companies for their security requirements.

This book will be of much interest to students of security, military studies and international relations in general.

Christopher Kinsey is a lecturer in the Defence Studies Department, King's College London, based at the Joint Services Command and Staff College. His research examines the impact of private security in international politics. He has published articles on the regulation and control of PSCs and more recently on their organisational structure.

CONTEMPORARY SECURITY STUDIES

CORPORATE SOLDIERS AND INTERNATIONAL SECURITY

The rise of private military companies

Christopher Kinsey

Routledge
Taylor & Francis Group

LONDON AND NEW YORK

First published 2006
by Routledge
2 Park Square, Milton Park, Abingdon, Oxon OX14 4RN

Simultaneously published in the USA and Canada
by Routledge
270 Madison Ave, New York, NY 10016

Routledge is an imprint of the Taylor & Francis Group, an informa business

© 2006 Christopher Kinsey

Typeset in Times by
HWA Text and Data Management, Tunbridge Wells
Printed and bound in Great Britain by
Biddles Ltd, King's Lynn

British Library Cataloguing in Publication Data
A catalogue record for this book is available from the British Library

Library of Congress Cataloging-in-Publication Data
Kinsey, Christopher
Corporate soldiers and international security: the rise of private military
companies / Christopher Kinsey.
p.   cm.   – (Contemporary security studies)
Includes bibliographical references and index.
1. Mercenary troops – History – 21st century.   2. Private security services
– History – 21st century.   3. Paramilitary forces – History – 21st century.
4. Security, international – History – 21st century.
I. Title   II. Series
U240.K516 2006
355.3′54-dc22   2005031184

ISBN10: 0–415–36583–X (hbk)
ISBN10: 0–203–01835–4 (ebk)

ISBN13: 978–0–415–36583–3 (hbk)
ISBN13: 978–0–203–01835–4 (ebk)

CONTENTS

ACKNOWLEDGEMENTS

When I began researching the role of private security in international relations I was under no illusions about how difficult and complex a systematic and comprehensive analysis of the subject would be. Writing about companies whose activities are frequently shrouded in secrecy and where fiction can be mistaken for fact or fact for fiction is no easy task. Nor has the task been made easier by the explosion in the growth of Private Military Companies (PMCs) in the last couple of years, especially since the start of the Iraq War.

Much of what has been written on PMCs either marginalises their impact on international peace and security, while pointing to the dangers posed by outsourcing military skills to companies that frequently work in a legal vacuum, or sees those individuals who work for them as white knights prepared to undertake peacekeeping operations that the international community would rather forget. Such opposing views do little to further understanding of this phenomenon. With this in mind, this book attempts to put forward a constructive argument which, while accepting that PMCs can make a positive contribution to international security, is highly critical of the way the industry is developing. The industry still has much to do if it is to attain the trust of the international community.

I would like to thank in particular the following people who were kind enough to share their views about the role of PMCs in international security: Dr Abdel-Fatau Musah, Brian Spratley, Ed Soyster, Lord Westbury, Andrew Kain, Tim Spicer, Lafras Luitingh, Nic van den Berg, Jim Hooper, Kevin James, Nigel Woof, Doug Brooks, and David Isenberg. Without their contribution, it would have been much harder to write this book. I owe them all a debt of thanks.

I owe a special debt of gratitude to Professor Michael Williams, who was responsible for supervising my doctoral research that resulted in this book, Professor Stuart Croft and Professor Martin Alexander, who both gave me constructive comments on what I needed to do to turn the thesis into a book, and Professor Colin McInnes, who gave much needed advice and encouragement throughout my research and without which this book may not have been written.

Michael Grunberg and Christopher Beese also deserve a special mention. For four years I bombarded Michael with questions about the nature of PMCs in general, but Sandline International in particular. Michael never tired from answering these questions, however insignificant they may have appeared to him, while his answers were always very lucid. Christopher was a source of invaluable information about ArmorGroup, in particular, and the industry in general. More importantly, he conveyed to me a very clear understanding of where the future of the industry lay and what companies needed to do to get there. And lastly, I owe a special thank you to my friend Mary Stenhouse for her untiring support and suggestions on how to improve various parts of the text.

Finally, trying to keep abreast of an industry rooted in secrecy and one that is changing by the day is an impossible task. Errors will have been made, for which I take full responsibility. That said, I hope I have done a reasonable job in producing an overall analytical framework to understanding the industry from a mainly British perspective. Furthermore, it is hoped the book goes some way to developing an underlining theory that will help us improve our knowledge of the role of PMCs in international security.

ABBREVIATIONS

ACOTA	Africa Contingency Operations Training and Assistance
ACRI	African Crisis Response Initiative
AECA	Arms Export Control Act
AFRC	Armed Force Revolutionary Council
BMATT	British Military Assistance Training Team
BP	British Petroleum
CCB	Civil Cooperation Bureau
CHOGM	Commonwealth Heads of Government Meeting
CIA	Central Intelligence Agency
CPA	Coalition Provisional Authority
DFID	Department for International Development
DOD	Department of Defense
DRC	Democratic Republic of Congo
DSC	Defence Systems Columbia
DSL	Defence Systems Ltd
DSS	Diplomatic Security Service
DTC	Defense Trade Controls
DTI	Department of Trade and Industry
ECOMOG	Economic Community of West African States Monitoring Group
ECOWAS	Economic Community of West African States
EO	Executive Outcomes
EOD	Explosive Ordinance Disposal
EU	European Union
FAA	Forças Armadas de Angola
FAPLA	Forças Armados de Popular de Libertaçac de Angola
FCO	Foreign and Commonwealth Office
FNLA	Front for the National Liberation of Angola
HMG	Her Majesty's Government
ICC	International Criminal Court
ICRC	International Committee of the Red Cross
IFI	International Financial Institution
IGO	Intergovernmental Organisation

IHL	International Humanitarian Law
IMF	International Monetary Fund
INGO	International Non Governmental Organisation
IT	Information Technology
ITAR	International Traffic in Arms Regulation
KBR	Kellogg Brown & Root
KMS	Keenie Meenie Services
MAG	Mines Advisory Group
MIC	Military Industrial Complex
MOD	Ministry of Defence
MPLA	Movement for the Liberation of Angola
MPRI	Military Professional Resources Inc
NATO	North Atlantic Treaty Organisation
NCO	Non Commissioned Officer
NGO	Non Governmental Organisation
NPFL	National Patriotic Front of Liberia
NPRC	National Provisional Ruling Council
OGEL	Open General Export Licence
PCC	Private Combat Company
PDD	Presidential Decision Directive
PMC	Private Military Company
PSC	Private Security Company
PVR	Premature Voluntary Release
RPF	Rwandan Patriotic Front
RUF	Revolutionary United Front
SADF	South African Defence Force
SAIC	Science Application International Corporation
SAS	Special Air Service
SBS	Special Boat Service
SLPP	Sierra Leone People's Party
SOE	Special Operations Executive
SSR	Security Sector Reform
UAV	Unmanned Aerial Vehicle
UN	United Nations
UNITA	Uniao Nacional para a Independencia Total do Angola
US	United States
USAID	US Agency for International Development

INTRODUCTION

Writing a book on Private Military Companies (PMCs) is not easy. The supply of military/security services is frequently shrouded in secrecy and, even though the industry is the focus of a fairly intensive media campaign at present, there is still a lot we do not know about the companies involved in the industry. Nor can we look to academia for help to understand what is happening within the industry. While there are some excellent books written by academics on the subject, they are few and far between. Moreover, if, as this book claims, PMCs are new actors in international security, academic research into them is essential to help policy makers decide how, or if, they should be utilised in support of humanitarian and national interests in an ever increasing hostile world. This book hopes to unravel some of the secrecy that surrounds the industry, while at the same time trying to explain the nature of this phenomenon.

The emergence of a new security actor

The international system has undergone significant changes since the end of the Cold War. None more so than in the area of international security. Previously, this had been the sole responsibility of state militaries. Governments are still reliant on their military forces to protect their borders and vital interests. But, with the end of the Cold War, they have started to turn for support to a new security actor, PMCs. Today, the international system is experiencing a huge increase in the number of PMCs operating on the international stage. They are in every respect global actors, operating on every Continent.

The global market for PMCs is mainly dominated by UK and US PMCs,[1] and this is why the book has focused on PMCs from these countries. In the case of the US, any research into the global market for PMCs that did not take account of how the market is dominated by US PMCs would simply not reflect reality. In effect, it is impossible to write about PMCs without including US PMCs since they are at the forefront of the changes to private global security. In this respect, they are market drivers according to Singer.[2] The reason for focusing on UK PMCs is that very little is known about them, what they do, the organisations they work for, who runs these companies and why. Yet, just like their US cousins, they too are global players. UK PMCs even dominate large parts of the private security

1

operations taking place in Iraq. Neither does their position, as leading providers of security to the country, appear to be weakening.

Accurate statistics which explain the nature of the PMC market do not really exist. This point was made in the UK Green Paper on PMCs and repeated by the House of Commons Foreign Affairs Committee report of session 2001–2. In each case, it was stated that obtaining information about PMCs abroad was hard and is often unreliable.[3] Furthermore, none of the witnesses who gave evidence to the Foreign Affairs Committee at the time was prepared to state very clearly the number of PMCs operating from the UK, though Tim Spicer, former Chief Executive Officer of Sandline International, estimated that there were about half a dozen.[4] Christopher Beese, Chief Administrative Officer for ArmorGroup, makes the same point. The problem with figures is the absence of definition: what is a PMC, what does it do, what services does it supply? Thus for ArmorGroup, trying to quantify the market is simply too difficult.[5] Moreover, the companies themselves are very sensitive about releasing commercial information into the public domain. This is not surprising considering that the business prefers to operate in a discreet manner, not advertising what they do. However, this is not to say that a picture detailing the nature of the market cannot be constructed; only, according to Mandel, that it is impossible to get a solid and comprehensive picture of the full scope and nature of security privatisation, let alone its impact within affected countries and on international relations.[6]

It is estimated that the PMC industry now generates US$100 billion in annual revenue and that PMCs operate in more than 50 countries.[7] In a list published in 2004 by the International Consortium of Investigative Journalists it was stated that, 'since 1994, the US Defence Department has entered into 3,061 contracts with 12 of the 24 US based PMCs'.[8] Singer points to the fact, that since the end of the Cold War, PMCs have been active in conflict zones throughout the entire world.[9] Furthermore, they have played a decisive role in several conflicts, their presence sometimes determining the outcome of the conflict. PMCs have carried out operations from as far afield as Angola to Croatia and Columbia to Papua New Guinea. They have operated on every continent barring Antarctica. ArmorGroup, for example, employs 4,000 people in 38 countries,[10] while Military Professional Resources Inc (MPRI) employs over 1,000 people and operates in every region of the world.[11] It is harder to gauge the extent of overseas operations of smaller PMCs. However, the government of Puntland, a quasi-state that emerged after Somalia fractured, contracted Hart Group to undertake coastal patrols on its behalf.[12] Olive Security, Erinys International, Rubicon, and Control Risks Group have secured contracts to provide security in Iraq.[13] Another PMC, Geolink, gave military support to Mobutu's regime but failed to prevent the regime's demise.[14] Such contracts are only the tip of the iceberg,[15] as the services they offer are integrated into the operating procedures of government, international organisations, and multinational companies.

Today, PMCs undertake a range of activities that hitherto would have been the responsibility of state militaries. Such activities can be divided into military operational support, military advice, logistical support, security services and crime prevention services. The last activity is not a military task but, because PMCs and private security companies (PSCs)[16] are able to draw on an extensive network of retired military and civilian police officers, providing the service seems only sensible. In the case of military operational support, PMCs guarantee to provide support for, or participate in, military operations for a government. Military advice involves providing training for state military forces, including Special Forces, covering weapons, tactics and force structure. Logistical support covers a whole range of services from supplying equipment, protecting humanitarian assets on the ground and helping to re-establish public infrastructure. Security services are directed at the commercial market and include guarding company assets and personnel. Finally, there is crime prevention. This area is also directed at the commercial market, but covers extortion and fraud.[17]

The market for private military security is set to grow.[18] Indeed, the role of PMCs in Iraq may be an indication that the use of such companies in stabilising foreign conflicts has become increasingly acceptable. The argument here is that it does not make sense to use a highly trained solider to guard a pipeline or bank. Forecasting market growth, however, is not so straightforward. Such companies make their profit in war, or by offering security to organisations operating in dangerous environments. However, predicting the next war, its intensity and nature, is very difficult. Even more difficult is measuring the degree of danger present in any given environment. Consequently, accurately forecasting market growth when the market is frequently changing is extremely difficult, if not impossible.

In the post-Cold War era, the enthusiasm for outsourcing government services has spread rapidly around the world. Linking it to efficiency and effectiveness, privatisation has been portrayed as a major step forward, its purported benefits contrasting sharply with the failures of over-centralised government bureaucracies. This approval for private-sector performance has resulted in many countries adopting this management system in virtually every conceivable sector, including the most fundamental of government functions: the provision of security.

One aspect of these developments that has raised significant interest, as well as controversy, is the PMC. Such organisations have been criticised by human rights groups who see them as no more than classic mercenaries providing quick-fix solutions to complex social and political problems of internal conflicts,[19] while others involved in the debate contend that there is a need to find ways to move beyond the polemical arguments on both sides of the security privatisation debate to try to make privatised security work, within the context of those carefully limited instances when it is employed.[20] Whichever side of the argument one is on, PMCs are now a fixture of international security that can no longer be ignored.

The issue of PMCs took on an immediate urgency for governments, especially the UK government after the UK-based company Sandline International became involved in Sierra Leone and the US company DynCorp's operations in the Balkans and Afghanistan. These operations were highly controversial. They raised serious questions about the right of PMCs to act in international security situations, and about the implications for the state of outsourcing its international security responsibilities to such actors. These debates have been further intensified by the events now taking place in Afghanistan and Iraq.

The privatised military industry: controversies and debates

The book examines the controversies and debates surrounding the emergence of PMCs on the international stage since the end of the Cold War. While some authors[21] see PMCs as nothing more than corporate mercenaries, others believe that such an assumption is too simplistic, arguing that it is not possible to understand them in absolute terms.[22] One of the contentions of the book is that they represent a new type of security actor on the international stage, different to the mercenary groups that plagued Africa in the 1960s and 1970s. Nevertheless, the roots of UK PMCs in particular are to be found in Africa during this period. It is important for policy makers to understand that the nature of private military security today is very different from the past.

Even though PMCs are relatively new to international security, they are increasingly being recognised by governments, civil societies and international organisations as legitimate actors that can have a positive impact on international security. Furthermore, the attitude of such governments as the UK and US is increasingly changing in favour of using PMCs to carry out certain security functions instead of using troops. The use of PMCs in Iraq by the US government is an indication of the extent of how far attitudes have changed towards the use of private violence. Even the UN Special Rapporteur on mercenaries has questioned whether security guards working for PMCs in Iraq can be defined as mercenaries.[23] Such changes in attitude are, in part, based on the assumption that PMCs are different to previous types of private violence. Moreover, while attitudes may be changing, PMCs still pose a real challenge for governments and international organisations.

Controlling the increasing impact they are having on international security will be a primary responsibility for governments, especially given the possibility that governments could find their monopoly on violence seriously eroded. The type of roles PMCs now undertake for governments and international organisations are not minor roles, but roles that can seriously impact on international security. Governments and international organisations need to broaden their understanding of the issues surrounding PMCs if they are to adopt appropriate policies towards them. To achieve this, they will have to engage with PMCs. At the same time this should ensure their impact on international security is not detrimental, while at the same time utilising the potential benefits PMCs can offer.

A major concern over the use of PMCs, and perhaps the most important, is the lack of a satisfactory legislative solution for the regulation of PMC activity. At present there appears no perfect solution to regulating PMCs while the interests of the companies, the government, and human rights organisations are so far apart. This is because no regulatory regime could meet the requirements of such diverging sets of interests. However, not to attempt to introduce new legislation to control the activities of PMCs is also unacceptable. Something must be done at the

Table 1   Sample list of companies providing private military/security services

Name	Country
Aegis Defence Services Limited <www.aegisdef.com>	UK
AirScan, Inc. <www.airsca.com>	US
AKE Group <www.akegroup.com>	UK
Alerta	Angola
Alpha 5	Angola
AMA Associates Limited	UK
ArmorGroup <www.armorgroup.com>	US
ATCO Frontec Corporation	Canada
Atlantic Intelligence	France
Aviation Development Corporation	US
Avient (Pvt) Ltd.	Zimbabwe
Beni Tal <www.beni-tal.co.il>	Israel
Betac Corporation	US
Blackwater Security Consulting <www.blackwatersecurity.com>	US
Booz Allen Hamilton, Inc. <www.boozallen.com>	US
Bridge Resources International	South Africa
COFRAS	France
Control Risks Group, Ltd. <www.crg.com>	UK
Corporate Trading International	South Africa
Cubic Corporation	US
Custer Battles <www.custerbattles.com>	US
Defence Systems Limited (ArmorGroup)	UK
Defense Security Training Service Corporation	US/Italy
DynCorp, Inc. <www.dyncorp.com>	US
Eagle Aviation Services & Technology, Inc.	US
Erinys International <www.erinysinternational.com>	South Africa
Euro Risques International Consultants SA	France
Executive Outcome, Inc. (confluita nella Northbridge)	US
Executive Outcomes	South Africa
Falconer Systems	South Africa
Global Contingency Projects Group	UK
Global Development Four, Ltd.	UK
Global Impact	Canada
Global Marine Security Systems	UK
Global Risk Management (UK) Ltd. <www.globalrisk.uk.com>	UK
Global Studies Group, Inc.	US
Global Options, LLC	US
Globe Risk Holdings, Inc.	Canada
Gormly International	US

continued…

Table 1 continued

Name	Country
Gurkha Security Guards, Ltd.	UK
HART GMSSCO Cyprus, Ltd. <www.hartgrouplimited.com>	UK
International Charter Incorporated of Oregon <www.icioregon.com>	US
International Defence and Security Resources NV	Belgium
International Port Services Training Group, Ltd.	Australia
International Security Consultants	Israel
ISEC Corporate Security <www.privatemilitarycompany.com>	UK
Janusian Security Risk Management Ltd. <www.riskadvisory.net>	UK
J & F Security, Ltd.	UK
Kellogg Brown & Root, Inc.	US
Kroll	US
Kulinda Security Ltd.	UK
MPRI <www.mpri.com>	US
Olive Security <www.olivesecurity.com>	UK
Presidium International Corporation	UK
Rubicon International Services <www.rubicon-international.com>	UK
Saladin Security	UK
Sandline International <www.sandline.com>	UK
SECOPEX <www.secopex.com>	France
The Risk Advisory Group	UK
Triple Canopy	US

Source(s): The majority of companies listed above can be found using their website address located next to the company name. Those companies without a website address may be found at <www. publicintegrity.org/bow/docs/bow_companies.xls> or by typing their name into a search engine.

national and international level. One starting point for the international community would be to establish a new international Convention directed at the activities of PMCs. Such a Convention though could take years to ratify. Alternatively, national legislation could be strengthened, particularly in countries that have a history of mercenary activity. What is clear is that any change to national and international law will also need to ensure transparency and accountability in the industry. Any regulatory regime that did not take account of these two issues would be greatly weakened and probably fail to maintain control over the industry.

Structure and sources

To attempt to present a well-balanced analysis of PMCs is a difficult undertaking, but even more so when reliable empirical information is limited. Those writing on the subject also tend to choose selective information that reinforces pre-existing prejudices. Such information that is presently available appears to suggest that the privatisation of security represents neither an overall advantage nor disadvantage to the international community, while one of the objects of the book has been to uncover positive and negative impacts of PMCs. To wait for a substantial increase in the amount of information on PMCs to be made available in the public domain is simply not possible given the pressing nature of the topic. Then again, all contemporary research faces this problem. To meet these challenges, a number of different sources, such as the internet, interviews, TV documentaries, official documents and newspaper articles were used to collect the information for the book.

The book relies extensively on interviews. Generally the interviews were very successful. Frank responses provided in-depth knowledge of individual companies and the industry in general. Furthermore, while interviewees were careful in what they said during an interview (what politicians call toeing the party line, or in this case the company line), they were willing to discuss controversial issues surrounding the subject that in turn helped to construct a comprehensive picture of the industry. For this approach to work, however, it was necessary to undertake such interviews with a certain degree of open mindedness about the industry to allow interviewees to open up. The book, however, does not rely extensively on interviews. Instead, the author attempted to use published sources wherever possible. Interviews were used as background information. They helped the author to gain a deeper understanding both of PMCs, and of the people who work for these companies.

The book also relies on a variety of primary documents. Sources include legislation, such as laws, *Hansard*, government and parliamentary reports, and private and commercial documents. These documents have been important in helping reveal political perceptions of PMCs. Here, the report by Sir Thomas Legg and Sir Robin Ibbs into the 'arms to Africa' affair, and the subsequent reports by the House of Commons Foreign Affairs Committee, and the government Green Paper on PMCs have all helped to shape ideas about potential uses for PMCs, as well as to explain how this particular event unravelled. Other primary documents analysed included policy documents from PMCs. These documents were particularly important in that they gave an insight into how PMCs understand their function as well as the limitations placed on that function.

Finally, the book is organised into six chapters. Chapter 1 lays out a set of general definitions of PMCs, before categorising the companies based on commonly agreed characteristics. Chapter 2 gives an historical overview of the marginalisation of legitimate non-state violence, while the last section discusses its re-emergence since the end of the Cold War. The third chapter discusses the evolution of private violence after the Cold War as an instrument of foreign policy. Chapter 4 then goes on to explain how the Sierra Leone affair caused tensions in this policy for the UK government. Following on from this, Chapters 5 and 6 examine the actual nature of the privatised military industry. These chapters focus on the role of PMCs in support of state militaries and the part they are now starting to play in the area of Security Sector Reform (SSR). Chapter 7 is concerned with the issue of regulation and control of the industry.

1

A TYPOLOGY OF
PRIVATE MILITARY/
SECURITY COMPANIES

Introduction

This chapter attempts to establish a typology of the industry based around the range and scale of activities private military and security companies undertake. In this respect, the key conceptual distinction lies in PMCs' resort to legitimate force: on whose behalf is that resort undertaken, and is it undertaken for the state, a public good, or for private gain? Based on commonly agreed characteristics of the companies, the chapter first attempts to categorise the industry by those working in this field. Here, the intention is to distinguish between the different types of private military and security companies, as well as individuals working as freelance operators. Next, an attempt is made to categorise the industry by examining the level of lethal force companies are either able or willing to project, and whether that projection has been taken in the public or private domain. This is achieved by looking at the intended purpose of that lethal force. Finally, by categorising companies, the chapter points to possible future impacts of PMCs and PSCs on international security.

There are problems associated with any attempt to categorise PMCs. In particular, the ability of individuals with a range of military skills to move between companies creates fluidity in the sector, increasing the range of activities and contracts a company can undertake. While some companies presume to be no more than a security company, they are able to undertake contracts normally associated with PMCs. As a consequence of this, it is easy for categories to become indistinct, especially when a company moves between governmental and commercial customers. Nonetheless, not to attempt to categorise companies will leave those who want to understand the nature of the business even more confused.

The first part of this chapter provides general definitions of the different categories companies fall into. These categories include private combat companies (PCCs),[1] private military companies (PMCs), private security companies (PSCs) and freelance operators. The inclusion of the type of PSC represented by Group 4 Falck, and large engineering companies that undertake PMC type activities is necessary to indicate the extent of the sector, although the focus of the book remains PMCs and PSCs engaged in quasi-military operations. The second part of the chapter looks at how the potential use of different levels of lethal force by different actors in the public, or private, domain can be used to categorise companies. In this respect, companies are plotted on two axes; the horizontal axis representing what is to be secured, while the vertical axis pertains to the level of lethality used to secure the object. The final part of the chapter focuses on a cross section of PMCs and PSCs, detailing their characteristics. The chapter discusses the position of these companies on one of the axes, explaining also the general problems associated with categorising the companies by using this method.

Problems of categorising PMCs and PSCs

The categorisation of types of companies in any industry, or of the correspondence between them and the realities they typify, is a difficult task, more so in an industry that sees companies able to undertake ranges of activities that cut across different categories of services, as well as moving between vastly different customers. This point has not been lost on those researching the private security industry. While there is broad agreement about the distinctions between mercenaries, PMCs, and PSCs, such labels are not necessarily helpful, particularly in relation to regulation. As Lilly notes, acceptable definitions are hard to find, and the different entities tend to merge into one another.[2] For example, how would we categorise a company such as ArmorGroup, able to supply military training and assistance, logistical support, security services, geopolitical risk analysis, and, if asked, crime prevention services through industry contacts, while the market for the company includes both commercial and government customers?[3] ArmorGroup is not the only company with such an operational range. Many, though not all, major companies possess this range. Smaller companies on the other hand, particularly in the UK, have access through networks to individuals that allow them to buy in the expertise to give them whatever range of activities they require for a given contract. Thus, while companies on the whole tend to stay within their areas of core competence, especially multinational security companies, this is not always the case when examining the activities of *ad hoc* security companies in particular. All security companies, however, have the flexibility to move between different categories of service if they feel it would benefit their financial position in the long term. As Lilly has argued, it is more useful to define the activities that companies engage in, rather than the companies themselves.[4] In reality, however, such movement between categories is unusual, especially when it comes to moving into the category of services covered by combat and combat support operations.

As Soyster points out, putting companies into categories is problematic, since companies often expand or contract their business or their focus.[5] As such, even though many companies appear settled, they will still resist being categorised, lest market opportunities are lost as a consequence. The best that can be achieved is to ensure categories are wide enough to cover the general nature of the work companies undertake. In this respect, categories simply differentiate between companies, such as Group 4 Falck, who can be said to be at one end of the continuum, and Sandline International, until it closed for business in April 2004,

9

which was at the other end. In the first example, Group 4 Falck's business concerns are primarily to do with commercial security, and entail only the minimum use of lethal force. In the second example, Sandline International, concern is with combat support operations for a government and can entail the use of very high levels of lethal force. Companies between both these extremes tend to see their activities run into each other.

At the same time, many field staff working in the industry tend to be multi-skilled. A single person, especially one retired from Special Forces, may have the necessary knowledge and skills to carry out more than one type of security task. Such individuals tend to move around the industry much more, covering a range of contracts from combat operations to commercial security protection. The criteria for them tend to revolve around money and whether they have the relevant skills necessary for a specific contract. Thus, the ability of individuals to cover a wide range of activities, while companies tend to remain within core competencies, has led to difficulties in constructing a typology of the industry.

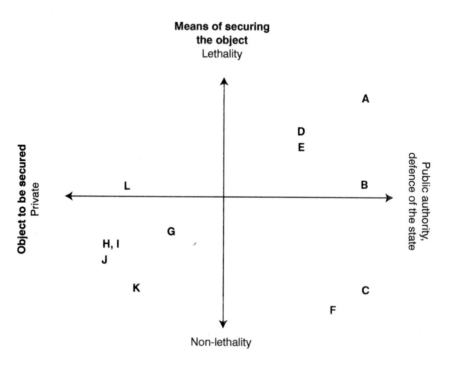

Figure 1   Locating private military and security companies by the object to be secured and the means of securing the object

The axes

The following axes are designed to help to categorise PMCs. The position of different PMCs on the axes is shown in Figure 1.

Defining the limits of the axes

The following section describes the international environment represented by the axes. The axes represent only the international environment. The reason for placing the axes within an international context is because the companies themselves operate globally. Thus the object to be secured may not necessarily be in the company's country of origin. Letters are used on the axes to represent both the companies and other organisations acting as controls, plotted on the axes. It is not intended for the axes to operate on a scale of numbers, using the number to determine the level of private and public interest and lethality. Instead, the companies are plotted in relation to each other, and three controlling agents as 'ideal types'. The controlling agents for the vertical axis are conventional police, paramilitary police, and the army. These controlling agents represent different levels of non-lethal and lethal means. The conventional police are seen as representing largely non-lethal means, the army maximum lethal means, while paramilitary police occupy the middle of the axis. Public private partnership represents the controlling agent for the horizontal axis, occupying the middle of the axis.

The horizontal axis represents the object to be secured. At the private end, the object represents anything from a private property, a commercial building, an oil refinery, to a mine, and so forth. At the other end, public authority is understood to mean the defence of the state. In this instance, the state is recognised as representing a legal territorial entity composed of a stable population and a government.[6] The vertical axis represents the means of securing the object represented on the horizontal axis. The bottom of the vertical axis is represented by non-lethal means employed by companies to meet their contractual obligations. Here, non-lethal is understood to mean that no lethal weapons are used to secure whatever the object being secured is. An unarmed guard of the type found in many modern shopping malls around the world comes very close to, or even occupies this area. Security is carried out without the use of any type of weapon, but is confined to reporting and potentially restraining transgression within a broader legal and police structure. The top end of the vertical axis is represented by lethal force, again employed by companies to meet their contractual obligations. Maximum lethality involves those techniques employed by an army fighting a war.

The axes provide four quadrants. In the bottom left-hand quadrant, security is a private commodity that can be purchased by anyone able to afford it. The security on offer is not deemed particularly violent, but, instead, can be described as relatively passive in response to the security offered in the top right-hand quadrant. Security companies working in this quadrant could be said to be working in a functioning social/political environment that enables the state to maintain stability. Moving up the line of lethality towards the top left-hand box would imply a deterioration of that environment as a consequence of the agencies of the state failing. The bottom right-hand quadrant represents an increase in state responsibility. In this quadrant, the state is deemed responsible for securing protection for society. At the extreme right-hand end of the horizontal line, no private security is available; security is the sole responsibility of the state. Companies working in this quadrant provide security services to the state. Security here represents a low level of lethality, and includes the supplying of private contractors as unarmed guards to government buildings or other public areas.

Moving into the top right-hand quadrant, state responsibility remains the same, while the level of security has increased to the strategic level. Here security is about interstate relations, while below it is more to do with domestic issues. The most prominent public institution responsible for security in this box is the military, able to exhibit maximum lethality under the control of the state. PMCs have also occupied this quadrant, although this is not a common occurrence, and has only really occurred in Africa. A reason for this may have to do with the inability of PMCs to project lethal force at the strategic level, while at the same time in support of the public good; Executive Outcomes' (EO) operations in Angola and Sierra Leone being exceptions. Certainly in the West, armies are large enough and technically capable to undertake the role of protecting the state. As mentioned above, this has not always been possible, especially in African countries. Here one PMC in particular was active. During the 1990s EO provided public security for Angola and Sierra Leone. In both cases, EO played a prominent part in securing the local government. Even after taking Iraq into consideration, it is unlikely, however, that the private sector will play a significant part in this quadrant for the time being. As the Green Paper pointed out, EO's success in Angola, and then Sierra Leone, may not be repeated.[7]

The final quadrant represents a political environment where state responsibility is in the process of breaking down, or has broken down completely, as in the case of Somalia. Here, security becomes a personal affair for the majority of local people. No longer is the individual able to rely on the state to secure themselves or their property, but instead has to either rely on themselves, local militias, or local warlords who are likely to levy some type of charge for this service. One could describe the political and social environment present at the extreme ends of the axes in this quadrant, as anarchy leading to chaos. Here, there is no government to govern the country, while violence is a factor in everyday life. In this respect, the quadrant is home to Kaldor's new wars, where people become social beings and not judicial subjects.[8]

Private security in this quadrant is undertaken by a range of security companies from ArmorGroup protecting assets owned by oil or mining companies, through small *ad hoc* PSCs employing freelance operators prepared to act offensively, to mercenaries out to exploit the social conditions found in this quadrant. Companies working in this quadrant have to accept that they may have to employ lethal force to fulfil the conditions of their contract, narrowing the field of potential security companies willing or able to undertake the security work in this quadrant. This in turn favours the employment of small *ad hoc* PSCs employing freelance operators with the range of skills to be able to operate successfully. These companies have tended to keep a low profile, drawing very little attention to themselves from the media, which their clients, especially the large multinational companies, find beneficial. Finally, this quadrant is also the home of the classic mercenary prepared to undertake most types of security work whether legal or illegal.

Distinguishing between companies and freelance operators

The following section reviews the main characteristics of military/security companies and freelance operators and the distinctions between them. A clear understanding of the types of companies involved in trading in military/security services is a necessary prerequisite to further understand the nature of the industry. Without it, it is not possible to map out the size and scope of the privatised military/security market, as well as the companies involved. The section first examines the hypothetical concept of a PCC, which is included because it helps us understand the extent to which war can be privatised. The fact that no PCCs exist at present has more to do with the lack of political will by states than with the ability of the market to establish such a company. Identifying the scope of the market for privatised war will enable us to separate out those military roles suitable only for state militaries, while allowing other military functions to be transferred to the private sector. The remaining sections are concerned with identifying the main characteristics of PMCs, proxy PMCs, PSCs and freelance operators. Unlike PCCs, such companies are a reality, and by separating them out in this way, we are better able to understand their purpose inside the security community.

Private combat companies

The purpose of a PCC is to highlight the extreme end of the spectrum. It is analytic and not, at present, real. Nic van den Berg,[9] along with other EO colleagues came up with the concept during the 1990s. The closest to the concept we had in the past was EO and Sandline International. Both PMCs were prepared to undertake combat support operations. Neither was it considered a problem moving from a concept to a position of actuality. As with PMCs, PCCs would be dependent on databases for their personnel. Thus, turning the idea into reality would simply require carrying out the same functions as those necessary to mobilising a PMC. Nic van den Berg describes a PCC as a PMC specialising at the sharp end of the security industry, which meant undertaking solely combat operations, leaving combat support and logistics to PMCs. Continuing, he explains that the idea behind a PCC is to assemble a fighting force capable of being deployed at very short notice into a combat zone, and if necessary, to suppress an aggressor with military force. In short, a PCC would enforce peace by means of military action. The principle objective of such a force would be to deter an aggressor, or potential aggressor, from achieving their stated aim. The force could also act as a buffer between warring parties, especially where the likelihood of fighting occurring is high. Finally, the force would have the capacity to rapidly deploy over considerable distances with its own supporting arms, possess tactical mobility, have dedicated lines of logistical support back to its home base, and have a clear mandate.[10]

The concept is useful because it shows the extent to which some in the industry have thought about the role of private violence in international security. For example, some believe PCCs can play a role in humanitarian intervention, while also putting forward plans showing how this would be done.[11] Moreover, the idea of a government or international organisation, such as the United Nations (UN), turning to a PCC for military help may not be so far off as first thought.

Private military companies

The international order first witnessed the emergence of a PMC in Angola in early 1993,[12] when EO was contracted to recapture Soyo, an area with considerable oil reserves, from Uniõ Nacional para a Independencia Total do Angola (UNITA). PMCs provide military expertise, including training and equipment, almost exclusively to weak or failing governments facing violent threats to their authority. They provide local forces, which may be poorly trained and lacking in military competence, with customised offensive capabilities that may have a strategic or operational advantage necessary to suppress non-state armed groups. According to Shearer,

> military companies are distinct from organisations operating in other areas of the security industry in that they are designed to have a strategic impact on the security and political environment of weak states facing a significant military threat.[13]

They do this by playing an active role alongside the client's force, acting as a force multiplier. They may even deploy their own personnel into a conflict, but under strict guidelines that sees such a deployment come within their client's chain of command.[14] In the above context, PMCs are understood by those in the industry to be service providers, delivering a business package that contains all the elements their clients require to retain the military advantage over rival forces.[15] Thus, PMCs argue that providing an integrated package sets them apart from arms dealers who only provide one part of a package, namely weapons.

PMCs are also permanent structures going beyond the need to simply satisfy a single contract. In this respect, they are registered corporate bodies with legal personalities, subject to legislation, and hired by governments, ostensibly to provide public security.[16] Since they are corporate bodies, they adopt business practices including the use of promotional literature, a vetting system for staff, and a doctrine, normally represented in the form of company policy or culture. They are also able to draw on the same support that all businesses can draw on, for example financial, legal, marketing, and administrative support. It might even be argued that the business sector represents the rationale behind PMCs. Indeed, according to Tim Spicer, 'PMCs are the official military transformed into the private sector in a business guise'.[17] It has streamlined the process by which legitimate bodies are able to call on an efficient fighting force able to achieve a strategic impact to resolve small, but intractable conflicts. The same point is brought out in Singer's analysis of how the industry is organised. It is not an overly capital-intensive sector. There is no need for heavy investment, or for extensive managerial apparatus, as is required to maintain a public military structure. Barriers to entry are relatively low, as are the economies of scale. Companies need only a modicum of financial capital to get started, but also to stay in business. Finally, labour input is relatively inexpensive, and, at present, widely available.[18]

As a corporate body, PMCs generally have a Board of Directors. The composition of these Boards generally includes former military officers, though whether they are accountable to shareholders depends on their position. For example, MPRI's corporate senior management always includes retired Generals.[19] The personnel structure of PMCs is not always clearly identifiable since the companies usually retain only a very small permanent staff to run the office and manage the contracts. The majority of their workforce is drawn from networks of ex-service personnel, whose details are held on a database. This approach has given rise to the term 'database militaries' when referring to PMCs. Subcontracting work in this way is common among a number of industries since it removes the cost of maintaining a large workforce when there are no contracts available. Such an approach does not necessarily lower standards, but can make vetting difficult; although PMCs are keen to stress the importance of having a stringent vetting procedure standardised for all licensed companies, in order to maintain high standards.[20] A further characteristic of PMCs is their ability to transform themselves from state to state,[21] establishing parent companies, or operating subsidiary companies.[22] This practice is also common among multinational organisations. In both cases, the reasons probably have to do with avoiding paying unnecessary taxation.

Proxy military companies

Proxy companies represent a subset of PMCs. A good example of a proxy company is MPRI. What makes MPRI different from other types of PMCs is the company's close working relationship with its own government, aligning itself with its government's defence policy. MPRI has not been involved in combat operations. The company does not allow its employees to be armed, and has even turned down offers, or at least queried security type work that requires armed personnel.[23] As such, the company cannot project lethal force of the type projected by EO, or some of the private security companies engaged in high-risk security operations in volatile areas of the world.[24] Instead, MPRI has taken on training roles, as well as giving military assistance to foreign governments on behalf of the US Defense Department.[25] The company provides the same service as would the US military, but at a purportedly cheaper price, and with a minimum political risk for the government, compared to a situation where it had to use US troops.[26] This relationship is an example of how a PMC can enable a government to attain its foreign policy objective without first having to secure congressional approval, and with the knowledge that if things go wrong the government can distance itself from involvement, though how far remains to be seen.

British companies have also sought a close working relationship with their government but none has characteristics of a proxy. For example, London-based Saladin Security has its employees, normally ex-British soldiers, working alongside serving British Army officers, seconded to train Oman government forces. 'A former SAS officer, now employed by a private company, noted in 1997 that he worked much as he did before but on the other side of the public/private fence.'[27] This type of close contact with the Ministry of Defence (MOD) and the Foreign and Commomwealth Office (FCO) is seen as essential to their business. Companies are also aware of the type of difficulties that would arise if their activities were in any way to conflict with government policy. The FCO do, in fact, have a list of companies willing to undertake work on the government's behalf, if the work is deemed too politically sensitive for its own military personnel to handle. A foreign government makes a request to the FCO, who can then give details of the companies able to handle the request.[28]

Private security companies

According to International Alert, PSCs have similar corporate characteristics and control structures to PMCs.[29] However, such a statement is only partly true, while confusion over these two groups is easy to understand given the fact that both groups are typically founded by former soldiers, carry guns, and adopt a tactical approach to their work. Yet each group has its own culture and doctrine and determines its own operational arena, while in terms of ownership and operations PSCs are little different to the commercial security companies operating within the UK.[30] Each of these groups maintain ethical policies, and attempt to interfere as little as possible in the political arena of the country it works in. Where major differences do exist between PMCs and PSCs is the range of services they provide. PSCs are generally concerned with crime prevention and public order.[31] The tasks they undertake range from countering fraud, risk assessment of insecure areas on behalf of companies evaluating investment prospects, armed guards to protect government and commercial installations and persons, and finally security advisers for multinational companies operating in the more volatile areas of the world.

The security activities of some PSCs have sometimes tended to move into the political–military area of activities occupied by PMCs. In the West, this can be attributed to a process of economic realignment between public and private interests, resulting in the transfer of military responsibilities or particular military functions to the commercial sector. In the developing world, it is sometimes attributed to the failure to maintain control over violence by the institutions of law and order. In this situation, as suggested by International Alert, the security activities of some PSCs have a bias towards more sophisticated security services associated with military security. This sophisticated approach is not surprising when we consider where these companies operate, normally in conflict prone countries such as Angola and Columbia. Maximising the use of technology to the extent that operations take on a military-like capacity is probably seen by the company as essential if they are to fulfil their contractual arrangements. At the same time such an approach can have a significant impact on a local conflict, changing the political and military environment to the advantage of local actors, for example local business entrepreneurs, and not necessarily the government.

Furthermore, a number of the roles mentioned above come very close, if not occasionally crossing over, from the commercial arena to the political military arena, as a consequence of the ability to have a limited strategic impact. Shearer has identified three examples of this:[32]

1 Security companies, together with international mining and mineral concerns, are allied increasingly more intimately to governments. Companies are frequently linked to powerful individuals within the state.
2 As instability in places such as Angola, Sierra Leone, and Columbia escalates, passive guarding activities become a springboard to more aggressive military actions in collusion with local companies and powerbrokers.
3 A more militaristic trend is the move by foreign security companies to train local police, militaries, or local companies to enhance security around key installations. This action entails the risk that there are no subsequent restrictions over how new found expertise might be used.

In relation to the first example, security companies in Angola have entered into joint ventures with senior figures in the Angolan military, with the intention of securing contracts to service the diamond mines and oil fields controlled by multinational companies. As Vines notes '[senior military figures] are paid hefty dollar salaries to sit on the Board of Directors of local companies.'[33] Lifeguard, staffed by former EO employees, has engaged in firefights with rebel factions in the course of their duties that have included guarding a number of strategic assets in Sierra Leone.[34] A more militaristic trend was supposed to have been demonstrated by Defence Systems Ltd (DSL). The company, it was claimed, was involved in training local forces, and was also accused of human rights abuses in Colombia, while protecting British Petroleum Exploration Colombia (BP) oil installations.[35] In April 1996, former employees of the company's subsidiary, Defence Systems Columbia (DSC), were accused by human rights organisations of involvement in killings, beatings, and arrests, while working for BP.[36] These claims have always been denied by BP.[37] The company did not welcome such publicity. As Nigel Woof, Vice President Marketing, ArmorGroup,[38] explained:

in practice many PSCs do conduct activities that some might consider military, for example, training police forces, or providing maintenance for military vehicles on peacekeeping operations. There is no clear line of what becomes 'military'. It is subjective and incremental. However, what [ArmorGroup] does is to always keep in mind the reputational implications of any new work [the company is] offered. [The company] knows [it] may ... be subject to some criticism, but in general [the company] aims to act in a way that is justifiable to [the company's] key informed stakeholders.[39]

The company AirScan, on the other hand, is an example of a US private security company that has carried out airborne security operations that are more akin to military style operations in such countries as Angola since 1995.[40] It should also be stressed that all these companies are relatively small compared to the commercial security companies such as Group 4 Securicor and Securitas. They also represent the borderline between the sectors, not fully engaging in military security operations, while, at the same time, not providing the type of security guards used to protect commercial premises in the West.

Finally, another area of operations PSCs are entering into is humanitarian support, which was once the exclusive preserve of charities. A number of security companies, along with other international organisations, have taken on demining operations in Kosovo for the British government. ArmorGroup is the lead agency in South West Kosovo and coordinates the activities of all humanitarian mine action organisations in the region.[41] The company was also shortlisted by the Department for International Development (DFID) to manage its crisis response in conflict situations.[42]

Commercial security companies

The inclusion of commercial security companies is an attempt to differentiate between PSCs and the larger publicly owned security companies that provide uniformed and normally unarmed guards for commercial, government, and international organisations around the world. The organisational structures of these companies exceed that of PMCs and PSCs. For example, Group 4 Securicor[43] is a multinational corporation that operates in 100 countries, with an annual turnover of about £3.80 billion and employs as many as 340,000 full and part-time employees.[44] They have also been able to use their size, corporate structure, and financial position to establish new roles for the security industry. Group 4 Securicor pioneered the private contracting of detention facilities and prisons in Britain, for example. The company is also involved in other areas of public concern around the world including immigration, fire services, and ambulance services as well as running rescue and safety courses for the maritime and offshore oil and gas industry. Furthermore, whereas 10 years ago the company would have looked at employing former soldiers, now the company is moving away from this approach for its services in developed countries.[45]

Freelance operators

This last group of actors, the freelance operators, is the most controversial, if in practice one of the smallest of all the groups, since many freelance operators closely resemble the classic mercenary. This is not surprising considering the general perception of what represents a mercenary. The traditional notion of a mercenary is 'a soldier willing to sell his military skills to the highest bidder, no matter what the cause'.[46] Mockler, on the other hand, believes the true mark of a mercenary is 'a devotion to war for its own sake. By this, the mercenary can be distinguished from the professional soldier whose mark is generally a devotion to the external trappings of the military profession rather than to the actual fighting'.[47] Others point to their lack of any ethical standards while Spicer argues that mercenaries are usually individuals, recruited for a specific military task. They have no permanent structure, no group cohesion, no doctrine and no vetting procedure.[48]

Again, it is not hard to understand why certain groups or organisations see freelance operators as mercenaries when we look at the relevant laws defining what a mercenary represents. The actual definition in international law as set out in Additional Protocol I to Article 47 of the Geneva Convention (1977) classifies a mercenary according to the following criteria:[49]

a. is specially recruited locally or abroad to fight in an armed conflict;
b. does, in fact, take part in activities;
c. is motivated to take part in hostilities essentially by the desire for private gain;
d. is neither a national of a party to the conflict nor a resident of a territory controlled by a party to the conflict;
e. is not a member of the armed forces of a party to the conflict;
f. has not been sent by a state, which is not a party to the conflict on official duty as a member of its armed forces.

The wording of the definition is such as to exclude those foreign nationals in the service of the armed forces of another country, as with those individuals that served in the International Brigades in the Spanish Civil War, and where the international community is willing to tolerate such persons from falling within the definition of mercenary. Article 47 of Protocol I also ignores foreign military personnel integrated into the armed forces of another state. Included here would be the French Foreign Legion, and the Gurkhas. The definition also leaves out those induced by ideology[50] or religion, and those who may not participate directly in the hostilities. Finally, those foreigners employed as advisors and trainers are also not included in the definition.

Such a narrow definition, where all six criteria must be met consecutively for a prosecution to go ahead has rendered Article 47 unworkable.[51] Indeed, to be convicted as a mercenary using the above definition would require the individual, or group of individuals, to be extremely unlucky.

The inadequacy of Article 47 to exclude mercenaries from supporting the armed forces of a foreign government has meant the line between them and freelance operators has become blurred where the activities undertaken by each group closely mirror each other. Under this definition, it is not hard to understand why freelance operators might be labelled mercenaries when employed on security contracts by PSCs outside of their country of origin. Abdel-Fatau Musah, for example, described Group 4 Falck as a mercenary outfit after one of their managers of South African origin was seized by the Revolutionary United Front (RUF).[52] It is not known whether the security manager was on a short-term or permanent contract with the company. Either way, some people considered the individual a mercenary because of the nature, and place, of his work; a foreign security officer guarding mines in Sierra Leone, fitted nicely their perception of a classic mercenary.[53]

Since the activities of freelance operators normally involve some type of participation in combat operations, the general public has a right to be sceptical about their motives. If circumventing the definition is so easy, the only alternative is to define freelance operators using a more appropriate definition. As mentioned above, the traditional notion of a mercenary is 'a soldier willing to sell his military skills to the highest bidder, no matter what the cause',[54] while Mockler believes a mercenary is 'devoted to war for its own sake'.[55] In each case, a freelance operator, or an employee of a company may still fall within such definitions.[56]

While the majority of people working in the security industry work for PSCs or PMCs, there are a number of individuals who work solely for themselves. This group operates either as freelance operators, or they set up a small business, acting as a gateway to a larger pool of retired military personnel through personal contacts. As one commentator explained, 'the firm ... basically consists of a retired military guy sitting in a room with a fax machine and a Rolodex.'[57] This appears to be one of the outcomes of South Africa's legislation to regulate PMCs.[58] As Nic van den Berg explains, 'the main aim of the legislation was to close EO down. This was the only course of action open to the government of the day because it did not know of any other way to handle the situation. So, rather than confront the government, or relocate to another country, EO decided to cease trading. This sent a signal to other companies or groups, and, as a consequence, many South Africans, including former EO personnel, now freelance for non-South African companies in Africa and elsewhere around the world'.[59]

Many freelance operators will have their name on more than one company's database, and if their range of skills permit will be able to move between contracts covering combat operations, manned guarding, to humanitarian demining. For instance, a former Special Forces operator with explosive ordnance disposal skills could work for a PSC carrying out demining work, for a PMC training foreign soldiers, or as a mercenary. This has led to a number of authors simplifying the terminology by referring to PMCs and PSCs as corporate mercenary firms.[60] Labelling the corporations as mercenary is not always intended to demonstrate revulsion towards such corporations, but points to the difficulty of trying to categorise both individuals and corporations involved in the business.

Whether all security companies would feel comfortable employing a freelance operator who may have worked in all three categories is questionable. Some of the larger security companies, with international reputations to uphold, would probably find this scenario unacceptable. It is unlikely that any reputable security company would want to take the potential risk involved. On the other hand, such persons allow a company to undertake activities not normally associated with their core competencies. Moreover, by employing freelance operators, an *ad hoc* security company could take on work associated with a PMC or PSC.

As the above argument explains, attempting to categorise both companies and individual freelance operators by the activities they undertake is problematic. In theory, companies can easily move between one of the above categories by buying in the expertise necessary to undertake contracts in that category. In practice, it becomes much harder because of the nature of the work associated with certain categories. For example, a large multinational security company is unlikely to want to move into the PMC category for fear of the impact this may have on its reputation, and thus future orders. *Ad hoc* security companies, on the other hand, may feel they have nothing to fear from being categorised. Indeed, because the type of work undertaken by *ad hoc* security companies closely resembles the type of work undertaken by PMCs, placing them in the same category could be seen as sensible, though whether the same level of expertise is to be found in *ad hoc* security companies as in PMCs is another matter.

A more useful way to categorise these companies may be to look at their resort to lethal force. Examining why companies would resort to lethal force, and the means by which a company would go about this, might allow us to better differentiate between companies. By doing so, we may be able to separate out the companies willing, but also able, to project lethal force at the strategic level in support of a sovereign government, from a company solely concerned with working in the much larger commercial sector, which arms employees for purposes of self-defence, as well as using unarmed guards to provide surveillance and alarm functions, for example parking guards. As the next section shows, the means of securing an objective for a sovereign state, compared to using lethal force for the purpose of self-defence could be said to be at opposite ends of a continuum, designed to plot the different level of lethal force companies are prepared to use.

Differentiating between private military and security companies using the object to be secured and the means by which the object is secured

In the previous section, an attempt was made to categorise companies by means of their characteristics. This, in turn, indicated the type of activities the companies were typically prepared to undertake. This approach indicates how companies may mutate to exploit different market opportunities. This is particularly so in the area of combat support operations, since companies can buy in the expertise *via* freelance operators, then dispense with their services when a contract is over. While such an approach may appeal to smaller *ad hoc* security companies intent on maximising their potential in the marketplace, large multinational security companies are less likely to want to jeopardise their reputation in this way.

While this categorisation goes some way to clarifying the sector, it is incomplete on its own. To supplement it, the chapter will now look at a second category: deployment of lethal force. Examining why and when companies would resort to using lethal force to secure their objective, and the level of lethal force they would use, allows the companies to be plotted on axes that, in turn, provides a fuller picture of the sector.

Index of actors

A   Traditional Westphalian state military[61]
B   Paramilitary police
C   Conventional police
D   EO
E   Sandline International
F   MPRI
G   DynCorp
H   ArmorGroup International PLC
I   Control Risks Group
J   Erinys
K   Group 4 Securicor
L   *Ad hoc* security companies

The actors

The following section describes the security actors that have been plotted on the axes. These companies constitute a cross-section of security actors operating globally. They provide a complete range of military and security services available to governments, Non Governmental Organisations (NGOs), and the commercial market. The inclusion of public security actors is to act as reference points for the latter group of private security actors.

Institutions of state security (A, B, C)

These organisations are all public institutions: they all have the ability to project lethal force, but at different levels. State militaries (A) are the ultimate fighting force.[62] Most armies in the world are able to project a higher level of lethal force than a PMC could ever be expected to wield. The fact that EO has projected higher levels of lethal force in Africa than some African armies is discussed below. A number of countries around the world operate a paramilitary police (B) alongside a conventional police force (C). Certain paramilitary police forces are also able to project significantly high levels of lethal force. For example, the Italian Caribineri and the French Gendarmerie operate within a military structure. They have also carried out peacekeeping duties in places such as Kosovo. This requires a willingness on the part of the paramilitary force to resort to higher levels of lethal force than might normally be expected in their home country. Conventional police may also carry sidearms, but these are only intended for self-protection, thus their projection of lethal force is seen as relatively minimal compared to the other two actors. As mentioned above, the reason for plotting these groups on the axes is to give an indication of the level of lethal force the actors mentioned below are able to project.

EO and Sandline International (D, E)

EO and Sandline International occupy the most controversial quadrant of the axes. EO mounted military operations where the level of lethal force was equal or beyond that of some state militaries in the developing world. Their military successes in Angola and Sierra Leone are evidence of this. At the same time, such operations were only in support of the local governments. The policy of Sandline International, which many saw as EO's sister company, was only to work for internationally recognised governments, preferably democratically elected.[63] The level of lethal force Sandline International could project is best gauged from the list of weapons it purchased for the Sierra Leone contract, although that contract is only a rough guide to the size and type of arsenal it could have assembled.[64]

While PMCs have tended to work for governments in developing countries, using force to secure a peace, some individuals now advocate they undertake certain military activities, notably to do with peacekeeping operations, for governments of developed countries, because of their ability to use lethal force successfully.[65] EO's use of lethal force in Angola and Sierra Leone has clearly demonstrated to the international community the capability of some types of PMCs in establishing peace in a war zone. Consequently, some in the international community see PMCs as offering an alternative to using state militaries in certain types of conflicts because of their ability to engage the enemy successfully. Thus, private military activities may increase in the quadrant occupied by EO and Sandline International, but only to the extent that the operations they undertake would be within their military capabilities, and sanctioned by the international community.

For this to happen, a swing towards outsourcing military tasks to PMCs would be required. This is already happening, though it is too early to say how successful outsourcing will be. There are a number of factors bringing about this change. They include shortage of manpower in state militaries, over-extension of military roles, especially after 11 September, and, in the case of the US, the problem of military personnel returning home in body bags. Overcoming these problems for any government will be difficult, and is likely to involve further outsourcing of certain military tasks in the future. DynCorp, for example, won a contract from the State Department to help support the Diplomatic Security Service (DSS) protect Afghanistan's President Hamid Karzi, while in Iraq security companies now protect convoys and installations from attacks by insurgencies.[66] If the trend continues, we can expect the same companies to carry out other types of roles that may include aspects of humanitarian and peace support operations using lethal force to achieve their objectives, while the level of lethal force they would be allowed to use is hard to gauge and would depend on the type of role they were asked to carry out.

MPRI (F)

The role of MPRI is somewhat different to other PMCs, especially those companies outside the US. MPRI's mission, as stated on the company's website, is to provide the highest quality education, training, organisational expertise, and leadership development around the world, while its main focus is on defence and public security development.[67] Its position on the axes closest to the end dominated by public authority and non-lethal use of force reflects the amount and nature of the work it undertakes for governments. A large part of the company's business is with the US government, much more than British PMCs undertake for the British government. The only other customers the company has are foreign governments. The company does not work for multinational organisations. They run a range of different types of military training programmes around the world for the US government designed to strengthen military and law enforcement institutions.[68] These programmes range from democracy transition assistance programmes in the Balkans to a programme called Africa Contingency Operations Training and Assistance (ACOTA). This programme succeeds the now defunct African Crisis Response Initiative (ACRI). Over the last five years, the ACRI programme has trained several thousand African soldiers in peacekeeping and crisis assistance tactics, while under ACOTA it is hoped to enable local countries to train their own troops.[69] The fact that the company offers training programmes designed to enhance the operational capacity of a state military, and does not act as a force multiplier, as other companies might, suggests why the position of the company on the vertical axis is close to the bottom. As discussed earlier, one of the distinctions between MPRI and Sandline is that MPRI will not allow its personnel to be armed, thus preventing them from entering a war zone. In this respect, they have been called a passive PMC, providing services up to and including training and support.[70]

A further, though less obvious, distinction, has to do with the type of former military personnel MPRI employs. Many of the company's programmes, though not all, are conducted at the strategic level. The aim of these programmes is to teach those responsible for a country's armed forces how to run them properly within the framework of a democratic system of government. MPRI claim to be able to perform any task or accomplish any mission requiring defence related expertise, military skills short of combat operations (or generalised skills acquired through military service), law enforcement expertise, and leadership development.

Thus, former officers and non commissioned officers (NCOs) employed to run these programmes would be qualified in these areas, and not necessarily in the skills used by Special Forces, and which PMCs are generally looking for in their employees. Finally, for MPRI to undertake training programmes at the strategic level, teaching other countries how to manage their military and law enforcement institutions, suggests that the client country has achieved a level of political stability necessary for the company to run programmes without the company being drawn into a potential civil war. For example, the company waited until 1995 before embarking on its Democratic Transition Assistance Program in Croatia. By this time, the local government had achieved the advantage in the civil war thus helping to stabilise the country's socio/political environment. Neither does MPRI take on contracts in countries facing civil war, leaving these contracts to other PMCs, or PSCs, prepared to take the risks associated with this environment. Indeed, the type of military skills MPRI are selling would be redundant in countries where the institutions of government have collapsed, as in Somalia. Here a PMC is more likely to be employed to physically participate on the ground in a security operation, though not necessarily involving combat.

DynCorp (G)

DynCorp's location on the axes reflects the company's position as a leading multi-national corporation in the field of technology. The company's core competencies are in the area of technological solutions to a range of industries, including the security industry. These include: information technology systems and solutions, technical services, international services,[71] and medical information services.[72]

It is the services supplied by DynCorp under international services that can be plotted on the axes. For example, DynCorp supplied former police officers to undertake law enforcement roles to the international police force in Bosnia, while the company has also undertaken contracts for the State Department in Colombia in its war against drugs.[73] This type of work is normally associated with PSCs, though the company does contest contracts associated with military training. While employees of the company working for the international police force in Bosnia would probably have to rely on a third party for protection, probably one of the intervening police forces in the country, reducing its reliance on lethal forces to a minimum, the company's role in Columbia is somewhat different. The company provides American pilots for herbicide fumigation and helicopter gunships that protect spray missions. They also provide search and rescue teams responsible for extracting crews from downed helicopters. In this instance, it may be necessary for the teams to engage hostile forces on the ground.[74] As such, the company has exhibited a level of lethal force higher than conventional security companies represented by Group 4 Securicor, but in a fairly restricted market and field of operation. The company has not as yet attempted to extend that level to a point comparable to EO's operations during the 1990s, while such a shift would involve moving outside their present core competencies. Thus, although the company is arguably a multifaceted company, its core competencies are in technical support, training, and logistical support, supplying armed guards, but not necessarily in the area of combat or combat support operations.

ArmorGroup International PLC (H)

ArmorGroup is an international company operating in over 30 countries around the world. The company was owned by Armor Holdings Inc, but was sold to a group of private investors led by Granville Baird Capital Partners of London in December 2003. The company became a public listed company on the London Stock Exchange in December 2004 and changed its name to ArmorGroup International PLC.[75] Armor Holdings acquired DSL in 1997 to form the basis of ArmorGroup. Today, ArmorGroup provides single source solutions to identify, reduce, and resolve exceptional risk in complex, sometimes hostile, environments. ArmorGroup's core competencies include:[76]

1 Information and business intelligence
2 Safeguarding brands
3 Countering fraud
4 Protecting operations in high-risk environments
5 Protecting individuals from kidnap threat
6 Humanitarian support
7 Integrated electronic security solutions
8 Computer security and forensic services.

While the UK government, in particular, is a substantial client of ArmorGroup, working for DFID on humanitarian support and landmine clearance operations, there is no indication that the company enjoys the same public/private partnership that some large technology companies such as DynCorp in the US, with interests in the private military sector, enjoy. This position may simply reflect the difference in the core competencies between ArmorGroup and large technology companies. The majority of public/private ventures in the defence sector are normally located within the technology/engineering sector, an area of competency the company lacks. Thus, ArmorGroup is positioned to the left of DynCorp, which is a technology company with considerable interests in long-term public sector defence projects that fall within the competencies of PSCs.

The activities of ArmorGroup plotted on the vertical axis have to do with protecting operations in high-risk environments, protecting individuals from kidnap threat, and humanitarian support. In each of these cases, there might be a requirement to use weapons. Even so, the company is very clear about when and how to resort to deadly force. As Nigel Woof explained, 'we arm our guards in some places where clients themselves need to be protected personally from armed criminals or terrorists. It is important to note that our rules of engagement stipulate that we would not use deadly force to defend property or other assets; only as a final counter measure against a potentially lethal attack directed against our own personnel or those of our client'.[77] This approach is fundamentally different to EO and Sandline International, which were willing to use deadly force for legitimate political purposes. Thus, it is assumed that, as a direct consequence of ArmorGroup's firearms policy, the level of deadly force would be substantially less than the amount Sandline International would have been prepared to use.

Control Risks Group (I)

Control Risks Group was founded in 1975 as a subsidiary of the then Hogg Robinson insurance and travel group. Today the company is a leading specialist in international business risk consultancy. The company has worked in 130 countries and has 17 offices worldwide. Prior to Iraq, the company's focus was the business market and protecting its clients from the risks, both physical and financial, found in certain countries. Working in Iraq has moved the company closer to the PMC/PSC market.[78] Control Risks Group services include:

1 Political and security risk analysis
2 Confidential investigations
3 Pre-employment screening
4 Security consultancy
5 Crisis management and response
6 Information security and investigation.

As with ArmorGroup the company does not have the same close working relationship with the government as some US companies have with their government. The company's position close to ArmorGroup on both the axes reflects the fact that these companies operate in the same market area as each other, even if each company specialises in one particular area of that market. For example, Control Risks Group has established a reputation for political analysis, while ArmorGroup is a leading company in humanitarian demining. However, both companies supply armed guards to the FCO and other government agencies. They have also supplied armed guards for some businesses working in Iraq.[79] Finally, as with Armorgroup, the level of lethal force the company can use is determined by the company's firearms policy, which restricts and controls those persons who can carry weapons, what weapons can be carried, and where they can be used.

Erinys International (J)

Erinys was formed in 2001. The company was set up in South Africa with the intention of focusing on the African security market. With the advent of the war in Iraq, the company made a strategic decision to go in on the coat tail of the coalition forces and establish business in the country. To achieve this they set up Erinys Iraq. Later on the company formed a holding company Erinys International, which now represents the central structure of the organisation.[80] The company is somewhat smaller than the majority of the companies discussed in this chapter, and is still a relatively new company to the security industry, establishing itself in 2001. Consequently, the company is still developing its business. The company offers a range of security services similar to those offered by ArmorGroup and Control Risks Group. Such services include:

1 Security services and consultancy
2 Emergency action planning and crisis management
3 Site security
4 Guard force management
5 Transportation and logistics security
6 Human resources.

As is clear from the list, the company shares the same market space as ArmorGroup and Control Risks Group, while the level and use of lethal force by the company is also similar. As John Holmes, a director of the company explains,

27

in Iraq we supply armed point security, which is defensive in nature, for the oil fields and close protection for individuals working in the country. The company also provides security for the US Corps of Engineers working in Iraq. In this respect, the company is no different from the majority of other armed guard firms operating in other areas of the world.[81]

Group 4 Securicor (K)

Group 4 Securicor[82] are representative of the mass commercial end of the security industry. Core competencies in this sector include cash collection, processing and delivery services, and manned guarding, including monitoring and response.[83] While the company represents commercial interests similar to ArmorGroup, over the last decade they have entered into public/private partnerships with numerous Western governments. This part of the security industry has taken on numerous state responsibilities, or gone into partnership with the state to help run public services.[84] These new responsibilities include managing prisons, prison transportation, education, immigration services, healthcare, and defence.

The position of these companies on the vertical axis reveals a cautious approach to contracts where the use of lethal force may be deemed necessary. Such companies are fully aware of the negative impact that a perception of them using lethal force might have on their business. Furthermore, it is now common for some of these companies to operate a strict weapons policy. One would also expect the policy to be in line with any ethical policy the company had introduced as with the case of Group 4 Securicor.[85] The company's resort to lethal force is only undertaken as a defensive measure. As a consequence of this, guards are only employed for their defensive skills, and the need for offensive skills of the type more closely associated with the military thus no longer exists.[86] In fact, the company has now started to move away from employing former soldiers on its contracts in developed countries. Where before this part of the security industry was dominated by military values, it is now looked upon as a civilian activity, with civilian values. Further enforcing this position is the company's effort to standardise its training and field operating procedures globally.[87]

As mentioned above, the resort to lethal force must be in line with the company's policy on the use of weapons. This is the company's position in countries where there is a serious threat to their field personnel, particularly those involved in moving large sums of money. Arming its field personnel is only common sense when the institutions of law and order have been weakened, and thus not able to provide even the minimum of protection. In this situation, the carrying of weapons can actually deter would-be attackers, increasing the level of safety for field personnel. Even so, the use of lethal force in these circumstances comes low down the vertical axis. This has to do with the choice of light weapons, the type of lethal force employed, and the reason for its deployment, primarily to protect the lives of field personnel in the execution of their duties. In this instance therefore, the role of armed field personnel is a far cry from the type of role carried out by those who work for PMCs prepared to conduct offensive military operations, as did EO in Angola and Sierra Leone.

Ad hoc *security companies (L)*

This group of security companies is by far the hardest to track. This may be because they tend to accept contracts involving questionable actions that border on the illegal, and therefore do not wish to draw attention to themselves. As a consequence of this, individuals working for this group are secretive of what they do so as not to find themselves, or their employer, in trouble with the authorities. A further effect of their questionable behaviour is the short lifespan these companies appear to have, sometimes lasting no longer than a single contract. The companies are easily set up, dismantled, or sold on to another company. Whether this is to hide the unsavoury nature of some of the activities they undertake, or simply a business practice to reduce costs, or both, is unclear.

Ad hoc security companies also have close links to some PMCs. For example, Lifeguard Management was a subsidiary of EO. The company undertook commercial contracts, while EO acted on behalf of governments. Lifeguard Management is now owned by Southern Cross Services.[88] Other companies include Alpha 5, Omega Support, Saracen International, and Stabilco.[89] The origins of these companies are not entirely clear. Again, it is possible the reasons for this are the same reasons mentioned above, though a number are believed to be South African.

While Lifeguard Management only undertook commercial contracts, other companies mentioned above have attempted to work for African governments directly, or for their officials. For example Saracen Uganda, another EO subsidiary, was set up to provide security protection for Branch Energy's gold mining operations. The company was established in co-operation with Major-General Salim Saleh, the half-brother of President Museveni; Saleh owns 45 per cent of Saracen Uganda.[90] Thus, as a consequence of them being prepared to work for both commercial and public interests, their position on the horizontal axis is problematic.

The position of *ad hoc* security companies on the vertical axis is an indication of their ability to resort to a level of lethal force not too different from the level of force a PMC can project. The difference in the levels of lethal force demonstrated by EO in Angola and Sierra Leone compared to the levels of lethal force associated with the companies listed above might simply reflect EO's ability to organise their operations better. In this respect, there seems little to choose between both groups in relation to the projection of lethal force. For example, the personnel of Lifeguard Management fought off an armed attack by an estimated 500 RUF rebels intent on destroying the facilities the company was guarding.[91]

Both groups also sought to challenge each other for the same security contracts during the early part of the civil war in the Democratic Republic of the Congo (DRC). EO and Stabilco even drew on the same pool of former soldiers for those contracts.[92] Even so, the position of *ad hoc* security companies on the horizontal axis is different to that of EO and Sandline International. This difference reflects a greater interest by *ad hoc* companies in private rather than state ventures. There are a number of reasons for this. First, as mentioned above, *ad hoc* security companies may not have the technical skills to take on government contracts, relying instead on the private sector for work. Second, the nature of the work in the private sector may not always be legally or ethically sound. In this case, some of the less savoury *ad hoc* security companies are ideal partners because of their flexible moral attitude as well as the speed with which they can be easily set up then dismantled once the contract is over, though it has to be remembered that this statement is a generalisation. As such, the position of individual *ad hoc* security companies on the horizontal axis may differ to the general position.

Explanation of movement around the axes

While human rights organisations, such as International Alert, express concern about the lack of regulation to stop companies from moving around the axes,[93] occupying areas in the quadrants that should be the sole responsibility of state militaries, in reality, the position of companies are fairly fixed. There are a number of reasons why companies find moving around the axes difficult. First, for well-established companies, entering a new market carries extra commercial risks they may not be prepared to take on board. For many companies, moving outside their core competencies could lead to them losing control over the way the company functions. This is particularly so if the company has to buy in the necessary expertise to enable it to operate in the new environment. For example, outside expertise may have very different ideas from the company about the way a contract should be handled in the new environment, and thus insist standard operating procedures be changed to facilitate these new ideas. For the company, this may mean reluctantly putting aside existing practices with a proven track record.

Not all such moves are necessarily successful. Thus, while companies in other industries may only have to absorb a financial loss, because of the nature of their work, security companies have to be especially careful lest the consequences for them of getting it wrong prove higher because of the emotional feeling the industry stirs, and the impact this can have on a company's reputation.

The second reason is that a company may not want to move too far outside its core competencies and the security environment that determines the type of contracts that are likely to be undertaken. Operating in a country where all political authority has broken down, and slipping into civil war, such as the DRC, might mean undertaking a security role on behalf of a failing government whose own army cannot undertake the role. This would require the deployment of a very high level of lethal force that would not necessarily be available to most companies, as well as unacceptable to some companies that do not see themselves as an alternative, or force multiplier, for a state military.

The third reason is simply financial and relates less to movement around the axes, than to movement within an established market. To undertake certain contracts could involve taking on large financial burdens. For example, running a prison is a very expensive business that involves a considerable financial outlay before a company is able to actually start operations. The operation, even if offset by guaranteed future revenues, and thus financing access, would exclude most, if not all, of the smaller security companies working within the domestic security industry, since such companies are unlikely to be able to raise the necessary finance needed to pay for the initial set up costs, or cover any contingency costs from problems that might arise. In the US, this problem is less acute because of the nature of the companies working in the area, namely large multinational corporations. Even so, such companies might still find it difficult to move outside of their core competencies.

Finally, some companies are closely associated with their government's foreign and defence policies. In this respect, US companies in particular carry out foreign policy directly, or operate within acceptable boundaries. Few act outside the national interests of their home government, or actively attempt to purposely embarrass their government.[94] Under these conditions, companies such as MPRI are unlikely to move into the area of the axes that represents the type of operations conducted by EO, lest they compromise their government's position, along with their own.[95] Some companies, on the other hand, may already be operating in this area. However, considering the social and political environment they must contend with at home, such companies are unlikely to operate openly in the way EO did, but, instead, keep a very low profile, while also ensuring their activities do not embarrass their government. Governments are also likely to ensure such activities from these companies cannot be traced back to them; that they can deny any involvement, even if that involvement only included passive approval. The subject is discussed in greater detail below. Finally then, while companies do move around the axes, the most respectable companies appear to stay within the areas of their core competencies, while there appears little evidence to suggest they would threaten their position in these areas by moving into new areas of business that might threaten their image and thus reputation.[96]

Conclusion

This chapter has attempted to establish a typology of the industry based on the range and scale of the activities undertaken by PMCs, PSCs, and freelance operators. The first part of the chapter described in detail the commonly agreed characteristics that differentiate between these groups and explores definitional problems. The inclusion of the hypothetical category of PCCs provided a useful benchmark against which to categorise existing PMCs, though it is unlikely the international order will witness the emergence of PCCs in the near future. Furthermore, as the PMCs and PSCs that operated in the early and mid-1990s demonstrate, distinctions based on characteristics reflect little more than the companies' own interpretation of their status. With the exception of a small number of PMCs, particularly in the UK and US, the characteristics of most PMCs reflect closely those of the *ad hoc* security companies. Distinctions that are drawn between reputable PMCs and *ad hoc* security companies relate to their core competencies, as is shown on the axes, and their customer base. In relation to each of these points, reputable PMCs are unlikely to move outside of their core competency unless accepting a contract that will enhance their reputation, while this in turn means working only for reputable customers.

A further attempt was also made to categorise companies using the axes to plot the level of lethal force they are either able or willing to project, as well as the purpose behind that lethal force, and its intended use in either the public or private domain. The axes are necessarily somewhat imprecise. However, they provide a useful framework within which the sector can be outlined and the operations of the companies categorised. Even so, it is very clear from looking at the positions of companies on the axes that there are wide differences between companies that need to be explained. Such differences relate to the understanding of the term security by each company. In simplistic terms, each company has a different understanding of which dimension of security they operate in. This understanding is articulated in the companies' core competencies, and this is what gives them their position on the axes. In the two top quadrants, security is associated with high levels of lethal force, while the actors can be both state and non-state. In the lower two quadrants, security is associated with low levels of lethal force, and again the actors can be state or non-state actors. Thus, while security in the top quadrants translates into the actual threat of using lethal force to protect self-interest, below, security typifies to a far wider understanding to do with notions of general well-being.

In each of these cases, categorisation of the groups presents problems. First, the transferable nature of the commonly agreed characteristics means distinguishing between each group is very difficult. Second, all three groups are able to undertake ranges of activities that cut across different categories of services, as well as moving between vastly different customer bases. Consequently, trying to define what we mean exactly by PMCs, other than a company that undertakes to provide military services for financial gain is problematic.

What does this tell us about the emergence of a different type of security actor? From the evidence above, there is no doubt that PMCs and PSCs represent a different type of security actor from the classic mercenary that operated during the 1960s and 1970s. Even though some PMCs are willing to project lethal force, unlike classic mercenaries, they have tended to restrict the use of such force to legitimate aims, only working for internationally acceptable governments and international organisations. Moreover, of the companies examined, only EO had projected the level of lethal force necessary to undertake combat operations of the type witnessed in Angola and Sierra Leone, while other PMCs have tended to stay away from this area of operations. The reason for this may have to do with them feeling uncomfortable undertaking such operations because of the political and economic risk involved, or they may simply lack the necessary skills. This situation though is likely to change in the not too distant future as the financial benefits start to outweigh the political and financial risk such operations pose.

Furthermore, by categorising PMCs and PSCs, the chapter has pointed to a set of future roles for PMCs and PSCs. Such roles as training African militaries to undertake peacekeeping operations can have a positive impact for international stability and security if properly managed. Close supervision by governments, however, will be needed to ensure PMCs and PSCs do not undertake roles beyond their capabilities. If such a situation was to occur it could have a negative impact on international security. Finally, it does look as if a different type of security actor is emerging on the international stage, with close ties to the state, but with its roots firmly fixed in the commercial sector. Why this security actor has emerged, and what it means for the future is the focus of the following chapters.

2

AN HISTORICAL OVERVIEW
OF PRIVATE VIOLENCE

Introduction

This chapter explains that the privatisation of military force in today's international order is not unique to recent times. At the same time, the chapter illustrates the different nature between previous examples of private military force and private military forces today. An examination of history from medieval times to the period of the Enlightenment reveals a strong role for private actors in military affairs. Then, mercenaries, privateers, and chartered companies were an integral part of the international order. Only with the arrival of the nation state, commencing with the French Revolution, did the general population question the right of these actors to exist. The chapter then provides a more detailed examination of the role of British mercenaries in the Cold War, focusing on their actions in the Yemen, the Hilton Assignment, and the Angolan debacle. Finally, the chapter examines the role of private violence in today's international system. Here, attention is given to examining the new development–security arena, and how the two areas of interest are converging. Violence is being privatised in this new arena as roles within the development–security arena are being allotted to the business sector. Consequently, strategic complexes are now made up of a range of different actors from the public and private sector. As discussed in Chapter six, two such commercial group are PMCs and PSCs, thus emphasising the revival of legitimate non-state violence.

The use of mercenaries in early European state transformation

According to Thomson, the marketisation and internationalisation of violence began with the commercialisation of war in Northern Italy as early as the eleventh century.[1] During this period, military power was truly marketised, democratised, and internationalised.[2] McNeill argues that this was due to the ease of transportation and communication experienced by city states in that part of Europe.[3] This allowed the ready importation of skills from adjacent, more developed areas, including the Ottoman Empire and the wider Muslim world. This importation of skills allowed the city states to rapidly expand their commercial interests throughout Europe

and as part of the process created extra wealth that allowed their inhabitants to Tilly
pay professional soldiers to fight their wars for them. Thus, there occurred a shift
in the social balance in Northern Italy that favoured merchant-capitalists over the
knights' relationship to feudal communities, which continued for approximately
200 years. Wars of this period were dominated, both in numbers and style, by the
Swiss pikemen or the Spanish *tercio*, as Europe continued to witness ever larger
concentrations of mercenary bands.

The knights' supremacy in war was dealt a mortal blow when, as McNeill
explains, 'an army of German knights met unexpected defeat in Northern Italy at
Legnano (1176) when they vainly charged pikemen who had been put in the field
by the leagued cities of Northern Italy'.[4] As a result of this defeat, there occurred
a shift in the military advantage towards this new form of warfare over its older
rival, as well as changes in the area of social leadership. Through the purchase of
professional military manpower[5] that the knights were unable to compete with
on the battlefield, townspeople were bringing to a close the age of the knight on
horseback, along with his ill-trained feudal levy.

The move to deploy pikemen was, in part, necessary if the tradable goods from
these cities were to be protected from marauding groups of armed bandits. At the
same time, such pikemen needed to be paid, notably by the city dwellers themselves.
The alternative was for townspeople to defend their surrounding countryside rather
than pay someone else to do so. Such a job would have required the townsmen
to sustain the discipline necessary to man city walls or to emulate the numerous
formations of pikemen in the field. The amount of individual commitment needed
to accomplish this necessary duty was now weakened as a result of the shift
in social bonds in favour of the market and the perceived benefits of a social
division of labour. Why would any person wish to take part in the defence of their
hometown when they could afford to pay some third party to do it for them? Thus,
Italian cities taxed their citizens to pay for the defence of their city. Even soldiers
were able to contribute to the cost of defending a city by spending their pay within
its walls, thereby putting monies back into circulation. As a result of these taxes,
argues McNeill, '[c]ities intensified the market exchanges that allowed [them] ...
to commercialise armed violence in the first place'.[6] The system was very much
self-sustaining, but with one major problem: how to go about producing a contract
that was acceptable to both parties and to ensure its enforcement.

The arrangement that finally emerged was for the civil administration to enter
into a contractual relationship with small units of professional soldiers. This
allowed civil administrators to keep better control over the city state's armed force
through the appointment of particular officers. The result of these changes was
an increased efficiency of the armed force by promoting those officers able to
function well, while marginalising those whose career advancement tended to
rely on more senior officers. There thus emerged a civil authority to oversee the
control of the armed forces, paid for by taxes. A corps of officers, whose career
advances depended more on civil officials with the power of appointment rather
than personal ties, also emerged. Thus we see in this arrangement both the idea
of commercialisation and the bureaucratisation of violence. These two important

changes to warfare were to spread north over the Alps to other kingdoms over the next two centuries.[7]

The wars of this period were, as Howard suggests, still medieval in their motivation; that is, 'they were fought to assert or to defend personal rights of property and succession, to reduce unruly vassals to obedience, to defend Christendom against the Turks, or the Church against heresy'.[8] But their frequency and intensity were now becoming politically and militarily significant as motivation for war started to shift towards the accumulation of territory. As state rulers accumulated territorial power, they had to be able to maintain their gains through the continued use of force. As a result of this development, a new order of power emerged, based on political, financial, and military muscle. This order was centred on sovereign princes, whose powers and rights were of a distinctive kind from those of counts and dukes. Instead of feudal rivalries, economic and military power came to dominate the relationship between monarchs. The consolidation of this power, in the hands of fewer and fewer monarchs, slowly put an end to private wars.[9]

At about the same time, a military revolution[10] had occurred that fundamentally altered the scale of war, and how it was conducted. The introduction of linear formations in battle, the ability to concentrate firepower more accurately, and the use of the momentum of the charging cavalry to break the enemy's formations, ushered in the modern art of war. The armies that carried through the military revolution were nearly all mercenary armies.[11] However, the transformation in the scale of war led inevitably to an increase in the authority of the state over military matters, while, at the same time, an increase in the financial burden of the state. Consequently, mercenary armies gradually became standing armies as the financial cost of implementing the changes introduced by the military revolution made the practice of disbanding and paying off mercenary armies at the end of each campaign season, and then re-enlisting for the new campaign season, extremely costly. It was now becoming financially viable for the state to transform mercenary armies into standing armies. Mass armies, strict discipline, the control of the state and the submergence of the individual, had arrived.[12]

Use of privateers

As far as sea power is concerned, the privatisation of violence is bound up with the practice of privateering. Privateering is defined in international law as 'vessels belonging to private owners, and sailing under a commission of war empowering the person to whom it is granted to carry on all forms of hostility which is permissible at sea by the usage of war'.[13] The owners of such vessels are normally required to place a bond to ensure compliance with their government's instructions. Their cargo is also subject to public inspection by warships. The act of privateering was thus a wartime act by which the state authorised private individuals to attack enemy commerce, and then allowed them to keep a portion of the prize seized as payment.

The origins of English privateering can be traced back to the eleventh century when the King ordered sailing boats of the Cinque Ports – Hastings, Hythe, Dover,

Sandwich, and Romney – to attack France.[14] In 1243, Henry III issued the first privateer commissions, which allowed for the King to receive half the proceeds of any prize taken by vessels authorised under the commission. English privateering appears to have had two golden ages. First, from 1544 to 1618, then from 1708 till about the middle of the nineteenth century when the major powers signed the Treaty of Paris on 16 April 1856 to ban the practice.[15] The decline in British naval capacity up to 1544 was finally reversed by Henry VIII when, in his war against France, he gave blanket authority for privateering. Previously, privateers were required to share their prizes with public officials, contributing to the decline of privateering.[16] As a consequence of these actions, his daughter, Elizabeth I, was able to gain naval superiority over the Spanish using private adventurers.[17]

The second golden age for English privateering began in 1708 when Parliament passed a new prize act. The new act allowed privateers to keep all the prizes they seized, as well as being paid a bounty by the government based on the number of prisoners taken. By the middle of the eighteenth century 'privateering had become something of a craze in England'.[18] There were even political lobbies formed to promote the interests of privateers. By 1803, privateering had taken on a new savagery and lawlessness unprecedented in modern maritime warfare, exacerbated by the inability of the governments of England and France to control the hordes of desperate privateers, and quasi-privateers, who were subject to their command.[19]

In England, privateers were auxiliary to the Royal Navy. The French approach to privateering was different. French privateers did not attack neutral commerce and were, in all respects, the French navy. For France, the golden age of privateering came after Colbert became Secretary of State in 1669. It was Colbert's enthusiasm for expanding French trade and building a navy that stimulated a heightened interest in maritime activity in general.[20] In any case, the principle threat to British trade in the wars between 1689 and 1815 came from privateers. As Crowhurst explains, 'French privateering was greatly stimulated by the wars between 1689 and 1713 which disrupted the lucrative trade in the Atlantic and Mediterranean, leaving merchants with little option but to invest in privateering.[21]

The determination not to allocate funds for naval expenditure exacerbated the French navy's financial weakness. Indeed, as a result of the collapse of the John Law's schemes in 1720, the French government was left without a central bank it could draw on for credit throughout the rest of the eighteenth century. In contrast, the English government was able to attain credit from the Bank of England. Such credit was essential because of the high cost of building and repairing ships. The lack of financial support from central government also fed into French strategic thinking about its navy. Since the main interest of the French navy was commercial, French merchants preferred a decentralised command structure at sea. This placed tactical decisions in the hands of captains who might decide on the course of action that was only advantageous to them. 'In time of war, therefore, the French commercial empire overseas was at the mercy of the British navy, whose ships acted in response to governmental decisions about when, where, and how they should act'.[22]

Just as with mercenary armies, privateering helped pay for the political transformation of the state at home, while the financial cost to the state of a shift to privateering was minimal. Privateering was seen as both an effective substitute helping provide resources for transforming the state, and a foundation for state naval power.[23] Privateering was essentially a weapon of weak states. They could deploy it against powerful states with little if any cost to themselves. However, as state navies grew, so the state's capacity to control the seas increased reducing the role of the privateer. As this section explains, the French and British both used privateering, while other countries also adopted privateering as a temporary alternative to building a state navy, demonstrating its importance as a feature of international relations right up to the middle of the nineteenth century.

Granting of commercial charters

The sixteenth century saw the creation of a new type of commercial entity, the mercantile company. Chartered by the state to engage in long distance trade and establish colonies, the mercantile company drove European imperialism for the next 350 years. They were unusual institutions in that they distorted the distinction between economics and politics, non-state and state, property rights and sovereignty, and the public and private.[24] As a consequence of these distortions, they presented their rulers with complex dilemmas.

Companies also varied from country to country. Dutch companies came closest to representing a purely private institution. French companies were really nothing more than state enterprises, while British companies fell between these two extremes.[25] Such differences reflected variations in the state structure of each country. France with its highly centralised state and weak commercial class saw the state take the lead in organising commercial adventures that were primarily concerned with increasing state power.[26] In Holland, where the commercial class took the lead in the hope of maximising profits, interference from the state did not occur until much later. However, in England the diffusion of power between the Monarch and Parliament produced a type of company that sought an intermediate approach to commerce that would satisfy both these groups.

Not only were companies given economic privileges and a monopoly over certain types of trade, they were also granted full sovereign powers.[27] Companies could raise both an army and a navy, build fortifications, make war, sign treaties, and had full legal power over their nationals.[28] The rationale behind companies being granted these extraordinary military powers had to do with the missions the companies undertook. In the case of the Dutch West India Company, they were specifically established to do as much damage to Spanish trade as possible by preying on Spanish shipping. The powers given them in their charter had therefore to reflect the offensive role the company was expected to undertake.[29] On the other hand, the Dutch East India Company was granted military powers to protect the company's trade monopoly and navigation east of the Cape of Good Hope and west of the Straits of Magellan.[30] Companies also had to protect themselves from attacks by local forces, pirates, or other Europeans. To do this, they built forts and garrisons. Such structures also acted as a coercive entry barrier, stopping outside groups from trading with local communities, while protecting the monopoly privileges of their owners.[31] Under such conditions, it was easy for a state to justify the granting of sovereign power on the grounds that, to establish a trading relationship with extra European areas, a company needed to be able to protect themselves, their ships, and goods from attack from other hostile groups interested in their business. Once the military infrastructure was in place, companies then argued that their trade monopoly was necessary otherwise they could not meet the cost of maintaining that infrastructure. In all probability, military power was used to impose and then defend their trade monopolies.[32]

The Dutch and English East India Companies enlisted Eastern as well as European mercenaries to protect their outposts from attack. In Surat, the English established regular infantry, cavalry, and artillery companies made up of Eurasians and Indians, including Indian mercenaries, to make up the numbers of the regular units when necessary.[33] The Dutch East India Company employed Indonesian mercenaries to help them establish their trading posts. Mercenaries were used to take Macassar, Sumatra, East Java, and Bantam in the latter part of the seventeenth century. The company also employed 5,000 European mercenaries and 20,000 local mercenaries to recapture its fort at Calcutta.[34] The hiring of foreign mercenaries in the East was not confined to the Dutch and English, but was an established business practice undertaken by everyone.

The companies were also prepared to use military force to confront local rulers who refused them the right to set up trading posts. One employee of the English East India Company, Henry Middleton, forced the Indians to trade with him by seizing Indian ships. After they had traded, he would then sell the ships back to their owners.[35] The Dutch used the same technique against the Chinese, seizing Junks to force the Chinese to trade with them rather than with Manila.[36] Not all military operations against local rulers were successful. This was especially so in India, where the company fought numerous wars throughout the eighteenth century.

In 1688, the English East India Company undertook military operations against the Moghul Governor of Surat in retaliation to harassment from local rulers. The Moghuls then retaliated by seizing company factories among other actions. In the end, the company was forced to concede, accepting a humiliating peace agreement.[37] The company continued to wage war on the continent throughout the eighteenth century. In 1757 they waged war against the ruler of Bengal. In southern India, the company fought four wars between 1769 and 1799 against the rulers of Mysore. Between 1775 and 1817, the company waged three wars with the Maranthas, before eventually establishing control over rulers. While the company was eventually successful, local resistance was not easy to dislodge. It was not until the third decade of the nineteenth century, by which time the British State was stepping in forcefully, that virtually the whole of the country south and west of the Punjab came under control of the English East India Company.[38]

The companies also turned their guns on each other. Most of the conflicts fought between the companies were over access to trade routes or markets. For example, the English companies fought their Dutch rivals for access to the spice trade, destroying each other's factories and ships in the process, while the English lay siege to the Dutch fort at Jakarta.[39] In Canada, the Montreal-based North West Company fought the Hudson Bay Company to break the latter's monopoly over the fur trade.[40] Such conflicts tended to coincide with wars between corresponding home states.[41] This is not surprising given the need for rulers to raise capital through the companies to fight their wars. Numerous conflicts were waged between French and British companies during the Seven Years War, while Dutch companies attacked Portuguese and Spanish trading posts in the East while the Dutch government was at war with these two countries.[42] At the same time, companies did not always comply with their sovereign's wishes in that they made war without permission from their sovereign if it suited their economic interests.

Mercantile companies were the creation of the state and yet were privately owned organisations. Their purpose was the pursuit of economic wealth. To achieve this, they used violence to first establish an economic bridgehead in a region, and then to impose their political will on that region. Without the military capabilities these organisations offered European rulers, it is doubtful whether they would have been able to exploit the markets in other regions of the world. More doubtful though, is whether these regions would have ever come under the control of European rulers.

The bureaucratisation of violence

All military activity requires organisation. For example, both the Roman armies and the crusaders needed to be organised into some type of fighting unit that could be supplied, as required, if they were to be successful in battle. The numerous functions carried out over large geographical areas required orderly accounting systems, specified chains of command and salaried civilian and military officials, capable of responding to these demands. The organisation of so many functions had thus to be standardised if any measure of success was to be guaranteed.

As monarchs achieved control over an increasing concentration of the means of coercion within their borders, so the need to improve the bureaucratisation of the military became apparent. Both concerns were, of course, opposite sides of the same coin. As military power became concentrated, so there was a need to increase bureaucracy that allowed for further concentration in military power. For legitimate non-state violence, the increasing pace of bureaucratisation of the military after 1560 assured its complete absorption into state militaries, or its removal from the battlefield.

The revolution in military affairs of 1560 established the need for the centralisation and bureaucratisation of both standing armies and navies. The reason for this lay in the tactical innovations introduced by Maurice of Orange and Gustav Adolf. There was now a requirement for troops to be properly trained and highly disciplined to carry out precise, complex movements as part of a formation in order to have a maximum effect on the enemy. Linear formation also lent itself to the practice of marching. This in turn favoured the wearing of uniforms. The adoption of a single uniform created a mass-psychological effect and that in turn helped enforce a single unified identity.[43] A physical 'as well as a psychological' environment had now been established, where the mass subordination of soldiers was possible.[44] The way was now clear for mercenary armies to be replaced by standing armies of the nineteenth century. However, the achievement of all these tasks required an army to have more than fighting skills.

The need for troops to achieve high levels of drill and training reinforced established convictions that mercenary armies were particularly well suited to carry through the military revolution.[45] These armies eventually became standing armies. As well as the need for uniforms and weapons to be standardised throughout units, the structure of a single system of rank at a general level in the army became necessary. The separation of functions was also further developed. The infantry and cavalry each had a clearly defined role on the battlefield, while improved bureaucratisation allowed for their support through the introduction or expansion of other important military institutions. Troops had to be supplied in the field, bureaucracy facilitated the simplification of this function. Those paid to fight fought, while their welfare became the concern and responsibility of others.

With the creation of a single identity within the military, soldiers now saw their units, from divisions down to platoons, as a surrogate family. The bonds between individual soldiers were in many cases stronger, because of shared experiences in war, than those between them and their real families. At the same time, such a unifying identity became more closely associated with the monarch. They wore the monarch's uniform, the monarch paid their wages, they fought for the monarch, but most important they swore an oath of allegiance to the monarch, all of which needed the bureaucrat to organise. The idea that a mercenary company could operate as efficiently without the support of a bureaucratic structure of this type is questionable. Thus we see, during the period 1560–1660, the amalgamation of mercenary companies into standing armies, while the introduction of innovative ideas in the area of tactics became standardised through a bureaucratic structure controlled by the monarch. This further reinforced the advantage of the standing army over mercenary companies. All of this was the result of the logic of efficiency necessary to enhance military power, the primary requirement needed to conduct successful war carried out between geographically diffused armed forces.

The era of industrial war

New weapons can change the way we think of war, as it was with the hydrogen bomb. But other inventions not associated with war have also altered our understanding of the way war is fought. It is this type of change that has had the most profound effect on the conduct of war in the last 150 years.[46] Wars up to the middle of the nineteenth century were still cumbersome affairs. Armies had certainly grown in size over the previous 200 years but their deployment in battle, along with all their supplies necessary to fight a campaign, was still restricted. This was because the means of getting armies to the battlefield still depended on those transport methods that had been established for centuries, notably the horse and cart, and the sail, the consequence of which was to place a limit on the size of war that could be fought. Improved transportation through the use of fossil fuels started to change all this.[47] Steamships and railroads were able to carry men, weapons, and supplies huge distances on an unprecedented scale. Now a European country was able to deliver the male population of fighting age to the battlefield, and keep it supplied there. As a result, countries started to count their soldiers by the million. No wonder then that the use of mercenary armies in war became irrelevant. States no longer needed the additional services of such groups of men, who in all probability would anyway be drawn into war as citizen soldiers.

The industrial revolution had made war into a series of specialised tasks involving hundreds of thousands of men and vast amounts of equipment, similar in many ways to what had occurred in factories. Just as production techniques were broken down into parts and formalised, so was the organisation of militaries in battle. Again the numbers of men needed to fight a war and the cost of the equipment they needed made mercenaries redundant. No longer could a commander control every aspect of the preparation for battle in his head. Artillery, infantry, cavalry, and engineers were now seen as individual parts that had to be brought together successfully if the battle was to be won. At the same time, the manner in which men and supplies were organised allowed for a constant flow of materials to the front line. In this respect, the industrialisation of war stretched beyond machine-made weapons to include the task of organising war. Only the state though, with its huge resources in manpower and production, had the capacity to undertake the task, leaving little or no opportunity for mercenary forces to participate.[48]

Bureaucracy, politics, and technology were now applied to military force, shaping its utility to benefit state elites. Bureaucracy organised men to fight more efficiently. Politics made war personal; nations by the start of the nineteenth century were being motivated to fight nations for reasons associated with national self-interest. Industrialisation increased the scale of war to an unprecedented level such that only nation states could raise the manpower to fight. This whole process was responsible for removing mercenary forces from state security from the middle of the nineteenth century to the end of the Cold War. The state had now come to dominate military power throughout the international system. There were of course individual acts of private military force, but these in no way threatened the state's monopoly over the control of force in the enforcement of international order.

The decline of private violence after 1856

The French Revolution (1789) saw a further transformation occur in the organisation of war. The rise of national armies and the marginalisation of the role of mercenary armies now occurred as a consequence of changes to do with sources of social power. Bureaucratisation, politicalisation, and industrialisation all came together to transform war in such a way that only a state with a political elite able to mobilise the country's entire population could take complete advantage of it. Even though total war was still some years away, the French Revolution marked the start of the rise of national state armies that required huge bureaucracies to run them, a national population willing to fight, and huge amounts of resources to sustain them in the field. In this environment, mercenaries were very soon marginalised as their usefulness to politicians and military commanders rapidly vanished.

While sizeable standing armies and navies were by this time common to the European scene, they were still regarded as the tool of the monarchy. This idea of war, understood as set moves on a chessboard, changed with the storming of the Bastille, igniting the French Revolution. War now became the prerogative of the people as they took control of those instruments necessary to wage it. Wars became wars between nations, fought by the citizens of those nations, as opposed to between monarchs with private armies. Clausewitz himself, one of the great nineteenth century military strategists, recognised this transformation in the organisation of war when, as Rapoport explains, he 'urged the replacement of cabinet wars by national wars, ... saying in effect, Give the War to the People! The State is the People!'[49] For the status of mercenaries, the implications were all too clear: there was neither a requirement nor a desire to employ mercenaries to serve the nation's interests. Since war was now identified with the pursuit of national interests, conducted with the 'full force of national energy',[50] the whole population of a country became involved. As Howard reminds us,

> By the end of the 19th century European society was militarised to a very remarkable degree. War was no longer considered a matter for a feudal ruling class or a small group of professionals, but one for the people as a whole. The armed forces were regarded, not as part of the royal household, but as the embodiment of the Nation.[51]

Nationalism enabled the state to centralise military power under its authority, while removing mercenary forces from the domestic scene; a task more or less completed by the beginning of the nineteenth century. States still used mercenaries, or privateers, operating beyond their borders to promote self-interest well into the nineteenth century. But state authority strictly controlled these groups.[52] Nationalism though was not the only reason for the demise of mercenaries from the domestic realms of the European states system. The industrialisation of war also sought to concentrate military power under the authority of the state. The huge destructive capability of modern weapons now made it necessary to draw on the male population if a state was to go to war.

The re-emergence of private violence on the international stage after 1945

The relationship between the British government and mercenaries continued throughout the Cold War. Shared political views and operational experiences between political elites in London and individual mercenaries, both of whom would have served in the British Army in the Second World War, increased the cohesion of the relationship.[53] This, in turn, made it easier for these two groups to establish informal networks with each other. This enabled the government to promote its foreign policy interests throughout the Cold War in places as far away as Africa, the Middle East, and the Far East.[54] As is discussed in more detail below, the best established companies undertaking covert operations were those employing former Special Air Service (SAS) personnel.[55] One reason for this is because they had close links to senior government officials and civil servants.[56] This type of intimacy not only encouraged networking through old contacts, but was well suited to operating informally. This approach ensured that the work of these companies did not conflict with government policy, but, instead, supported the interests of the country. At the same time, MI6 was monitoring all this activity.[57]

There are no better places to observe these networks operating than in Africa and the Middle East after the Second World War. Both of these regions had been dominated by British foreign policy before the war. But, by 1945, Britain was in decline, struggling to hold on to its empire. By the late 1950s a second and more far-reaching wave of decolonisation was under way that was to complicate Britain's interests in both these regions. A good example was the mercenary operation by the UK government in Yemen, where mercenaries were used instead of regular British Army troops to maintain an influence in the region. By inserting a small private force of retired Special Forces personnel, with the ability to have a strategic impact on the war, the British government was able to influence the direction the war took.[58] Indeed, as is discussed below, the idea behind the modern day PMC developed out of the country's involvement in the civil war in Yemen.

As recently as the 1980s, according to Bloch and Fitzgerald, mercenaries were preferred if the British government was to support an insurgency. As such, they were, and still are, subjected to relatively tight political scrutiny, and those operations they try to engage in that run counter to official foreign policy are blocked. Bloch and Fitzgerald argue that the reason for this is to do with the British government's sensitivity to allegations of subversion, and carefully trying to preserve its international reputation.[59] This approach has allowed some initiatives to be discreetly promoted by Whitehall mandarins, since, in the event of something going wrong, they are completely deniable, again keeping the secret world separate from liberal Britain.

Yemen

Britain's commercial interests in the Middle East were critical to the country's financial security.[60] As Bower explains, the military base in Aden had become a regional bastion of British military power, while the construction of a refinery in 1954 allowed the Royal Navy to command the approaches to the Red Sea and the Suez Canal, as well as claim a strategic presence in the Indian Ocean. Thus, to secure Aden's future, in February 1959, the Macmillan government established the South Arabian Federation.[61] At the same time, the country's influence in the region had been dealt a severe blow when the Suez debacle forced Britain to recognise that it was no longer able to assert control over countries in the region, as it had done before the war. With help from the Soviets, Nasser started to undermine Britain's position in the region. Then, on 26 September 1962, a group of leftwing army officers, inspired by Nasserism, led a *coup d'état* against the then ruler of Yemen, Imam Mohammed al-Badr, to establish a People's Republic. The situation in Yemen was of great concern to the British government. Egyptian support for the revolutionaries might encourage them to carry the fight to the South, thus threatening Aden and the South Arabian Federation.

During a visit to Britain, King Hussein of Jordan urged Julian Amery,[62] the then Minister of Aviation, not to let the British government recognise the Republicans, since 'Nasser just wants to grab Saudi Arabia's oil … .'[63] Both men agreed that Neil 'Billy' McLean, should now be dispatched on a tour of the area so that he could deliver an informed report of the situation to Prime Minister Macmillan. During his tour, McLean cabled Amery to report that the government was neither in control of all of the country, nor recognised as the legitimate authority by the tribes that supported the Imam. It was therefore wrong for the British government to recognise a new government in Yemen.[64] McLean again reiterated the position of the Royalists during a meeting he had at White's[65] with Sir Alec Douglas-Home, the then Secretary of State for Foreign Affairs, Amery, and Stirling, who had been known to Amery and McLean since their war days in Cairo. It was at this meeting that Sir Alec Douglas-Home, the most senior person present, Amery, and McLean decided something had to be done unofficially if Britain's position in the region was not to be further compromised.

Macmillan, now coming under increasing pressure from US President John F. Kennedy to recognise the new regime backed by Nasser, rejected this initial report by McLean. Kennedy was intent on building bridges with Nasser, but recognised this would not be possible if the Egyptian president was forced to pull his troops out of Yemen.[66] Then, on 19 December, at a meeting in Downing Street, McLean finally persuaded Macmillan that the Americans had misunderstood the situation in Yemen, and that Nasser's subversion could be halted. 'It was, [in Amery's opinion] one of the few turning points in history, which [he] had witnessed'.[67] At the same time, Macmillan was determined not to annoy Nasser and Kennedy. Thus, instead of giving direct assistance to the Royalists, support would be funnelled through unofficial channels.[68] To undertake this assignment, he turned to Stirling, the founder of the SAS. Working with McLean, the two men established an office in London to recruit former SAS officers as mercenaries to assist the Royalists against Nasser. Meanwhile, Macmillan took a secret decision to appoint Amery as Minister for Yemen, placing him in overall command of the operation. This decision was taken to prevent any cabinet member from obstructing a policy designed to support British interests. It also allowed Macmillan to place some distance between him and the operation, as Amery would take responsibility if anything went wrong.

The aim of the operation was to properly organise a resistance movement from within the Royalists prepared to fight. Few Royalists had any military experience let alone experience of weapons. Many had not even seen a machine gun at close quarters. What the Royalists wanted was professional weapons, communications training, and medical training.[69] In the case of medical training, rendering medical assistance in a country that spurned medicine, preferring instead to trust Allah, was particularly important to help gain the support of the Royalist villages. The operation entailed training the Royalists in the use of mines, mountain and desert tactics, machine guns and mortars. The mercenaries were also to collect and transmit all the intelligence possible on the Egyptian disposition, tactics and intentions. Finally, and probably their most important role, was to be the motivating power behind the Royalist efforts.[70]

The impact of the mercenaries on the war was dramatic.[71] The original Egyptian force sent to the Yemen by Nasser was estimated to number approximately 12,000 in strength. When the Royalists started fighting back, this number increased to about 30,000 in order to protect lines of communication and airfields. However, after the intervention of the mercenaries who immediately set about coordinating Royalist attacks on Egyptian forces, Egyptian numbers increased to an estimated 52,000.[72] Despite the atrocious reprisals carried out against Royalist villages, the Royalists, supported by the mercenaries, continued to mine, ambush, and snipe at Egyptian forces and the Yemeni Republican Army, which had placed great faith in the might of the Egyptian Army. Four and a half years after the start of the operation, 68,000 Egyptian troops were still in the Yemen, trying to extract themselves from the mountains. Consequently, they were in no state to interfere with the British Army's withdrawal from Aden, or pose a threat to the Saudis in the north.[73]

No one should be surprised at Macmillan's decision to support the Royalists through unofficial channels. British interests in the region were under threat.[74] Yet Macmillan did not want a repeat of the Suez debacle by upsetting the Americans.[75] On the other hand, Britain had the ideal tool for undertaking such an operation in former SAS personnel. This group in particular understood how important this region was to the country's interests. Amery, McLean, and Stirling all had considerable military experience, while Stirling and Amery were well versed in guerrilla tactics. Both men were also able to call on former wartime colleagues for help, again with experience in guerrilla warfare. Two men in particular were sought out: Colonel David Smiley and Colonel Brian Franks, the driving forces behind the reformation of the post-war SAS.[76] This group in turn were able to draw on retired SAS members who could be trusted. There now started to emerge a homogenous network of individuals well known to each other, who were either part of the official establishment or private individuals with strong links to it.

Watchguard

It was Macmillan's decision to act unofficially to defend British interests in Yemen that appears to have placed the idea of forming a commercial company to operate in areas of national interests in Stirling's mind. The result was Watchguard, the first commercial company of its type selling military style services to be set up in Britain after the Second World War and the first PMC to be established. It was also the first attempt by individual persons exporting military training to formalise the existing relationship between them and the British government. The idea behind the company was to safeguard British interests in places where the government was not able to act for whatever reason. As Hoe explains,

> The company would provide a service aimed at preventing the violent overthrow of a government, but it would not thereafter seek to exert political influence; indeed as a wholly independent commercial company it would be incapable of doing so. The company would not accept as a client any government that consisted of a racial minority; no client that was or looked set to be hostile to HMG [Her Majesty's Government] would be considered. The company would offer only instructional training and advice.[77]

The nature of the relationship was also different. The company sought to formalise its relationship with the government along business lines. As a consequence of this, the company took the decision not to feel constrained to refuse a task if FCO objections were unreasonable.[78] Watchguard International was finally formalised in the Channel Islands in 1967.

The spearhead of its operation was to be the Middle East, while the company's director of operations also identified Africa as a potential market. The company was to provide a range of services to Third World countries, particularly in the two regions mentioned above, which would support, or further, British interests in these regions. These services included:[79]

- Military survey and advice.
- Undertaking security for heads of state friendly to the British government.
- Training Special Forces.

Importantly, none of these services was undertaken without the consent of the British government. Stirling himself recognised the need for the government to consent to a commercial operation selling military skills that could run counter to British interests later on. In a private interview given in 1979, he confirmed that Watchguard was blessed with complete government approval:

> The organisation was designed to tackle really important military objectives which could not be tackled officially because of questions in the House of Commons. The British government wanted a reliable organisation without any direct identification. They wanted bodyguards trained for rulers they wanted to see survive.[80]

As Bloch and Fitzgerald explain, Stirling's statement unambiguously confirmed that private enterprise was used by government to broaden its range of covert action options beyond those it had with its own agencies in support of foreign policy.[81] At the same time, such help from the commercial sector allowed the government to circumvent the vagaries of parliamentary oversight.

By 1970, Stirling's original idea of placing Watchguard at the government's disposal had all but disappeared, overtaken by commercial interests.[82] The company now shifted its interests over to the commercial arena, placing profit before patriotism. It still sought to work closely with the British government but not in the way Stirling had imagined. [83] It is probable that the formal nature the company was now adopting did not sit comfortably with the way the government of the day preferred to conduct this type of business, in an informal atmosphere with the knowledge that if anything went wrong they could deny having played any part. By formalising the relationship, the company removed its one great asset for the government, the ability to deny having anything to do with them. In the end, this shift in the relationship caused too many tensions that could not be resolved. These tensions were to surface during one of Stirling's operations, which became known as the Hilton Assignment.

The Hilton Assignment

On 1 September 1969, a 28-year-old army officer, Muammar al-Gaddafi, led a group of young Free Unionist Officers in a bloodless coup against King Idris of Libya. In February the following year, a counter coup was launched from Chad led by a member of the Royal Family and a group of *émigrés*. The counter coup failed. Then in May that year MI6 launched a major covert operation, designed to topple the regime because it opposed British interests.[84] This became known as the Hilton Assignment.[85] The operation was to be one of MI6's last attempts at overthrowing a regime.[86] Again, just as in the Yemen, the need to conduct the operation at arm's length while under the tight control of officials from MI6 was paramount. The government could not afford the operation to be traced back to them. Therefore, the decision was taken to use French mercenaries.[87]

Initial contact between these two groups took place on 18 May when an unidentified retired high-ranking official, who had previously served in Libya, met Stirling in London.[88] Also present was a former MI6 officer, Denys Rowley, alias James Kent.[89] The trio discussed the Libyan government's closure of all British and American bases, and the nationalisation of BP oilfields. With two former SAS men to help recruit for the operation and MI6 organising the arms through an end-user certificate made out to a dealer in Chad, the organisation of the operation initially seemed to take on a homogenous character similar to the operation in Yemen.

The operation was initially set for November but was called off when the Yugoslav authorities confiscated the entire consignment of arms and explosives. After several false starts, the operation was rescheduled for the following February, but on 29 December MI6 finally cancelled the operation on the grounds that its secrecy had been compromised. A colleague of Stirling, Jim Johnson was approached by MI6 who told him that it was now common knowledge that French mercenaries were going to carry out the operation, while the name of the boat from which the operation was to be launched was also known. MI6 now asked Johnson to go to France to meet with the mercenary leader Roger Falques, who was also aware of the loss of secrecy and who was now convinced the operation would end in a complete fiasco, and no longer wanted to have any part in it.[90] To Stirling, it now seemed that MI6 had conspired against him. Shortly afterwards, he quietly withdrew from the mercenary business, a decision that probably had as much to do with the changes that were going on inside the business as it did with the failure to launch the Hilton Assignment.

British mercenary involvement in the Civil War in Angola

The trigger for the Civil War in Angola was the surprise coup in Portugal in April 1974 when a group of left-wing army officers ousted the Caetano regime. This was followed by a decision to withdraw from Angola after transferring the country to majority rule. In January 1975, the three parties contending for power signed an agreement in Alvor in southern Portugal that set in place a tripartite transitional government. The agreement provided for the withdrawal of Portuguese troops by April, to be replaced by an integrated armed force drawn from all three factions.[91] The largest and best organised of the three parties was the Marxist–Leninist Movement for the Liberation of Angola (MPLA), which threatened to dominate the transitional government as well as any post-independent government. Led by Dr Agostinho Neto, and drawing support from eastern and central areas of the country, the MPLA now became the target of the two other parties, who had also fought the Portuguese for independence.[92] The Front for the National Liberation of Angola (FNLA) controlled parts of the North, was headed by Holden Roberto, and received support from the American government. UNITA operated in the South under the leadership of Dr Jonas Savimbi, and received support from the South African and American governments. What set this civil war apart from the civil war fought in Yemen a decade earlier was the context in which it was fought. Unlike Yemen, Angola served no British interests. Instead, Angola was a war over competing ideologies dominated by the superpowers.

Since the FNLA received the majority of its financial support from the Central Intelligence Agency (CIA) it was only natural the CIA also conducted the covert military operations undertaken by British mercenaries, a decision the CIA would later regret, since their agents were not trained as military advisers.[93] Early on in the war, the 40 Committee[94] had approved US$300,000 to support forces opposing the MPLA. This amount was to increase until the total budget reached US$25 million by August 1975.[95] Permission to recruit mercenaries in Britain was given to the CIA by the British government, but only after the government had turned down at least two formal approaches for Britain to ship missiles to the FNLA. The government did not want to undermine its public position of concerned neutrality.[96] But, by allowing the CIA to recruit British mercenaries *via* an outside agency with no contacts into the establishment, MI6 officials effectively lost control over who was recruited and how they conducted themselves.

More important, the main concern of the mercenaries recruited was monetary, while many of them lacked the professional standards associated with soldiering. The decision to allow Dr Belford, who was at the time working for the FNLA, to recruit Costas Georgious, otherwise known as Callan,[97] bore this out. Initially hired as a medical orderly to assist in the FNLA military hospital at Carmona,[98] Callan quickly became involved in the actual fighting. He also became involved in organising the recruitment of further British mercenaries. Unfortunately, while Callan proved capable of daring deeds on the battlefield, he lacked the leadership skills and military training that Amery's group had, and that were necessary to conduct an insurgency operation. Neither was he able to construct a group identity,[99] as had been possible with those who had operated in Yemen. Furthermore, many of the mercenaries recruited also lacked the necessary insurgency skills associated with this type of operation, while their sole purpose for being in Angola appears to have been the lure of money.

The changing nature of private violence during the 1970s

At the same time that British mercenaries were operating in Angola, private military security started to emerge as a legitimate activity. Changes to the international environment in the mid-1970s created a new set of social conditions that in turn encouraged the legitimate growth of the security industry. According to Lord Westbury, globalisation and international terrorism created opportunities for companies to engage in private military security as a legitimate commercial activity.[100] Multinational corporations, in particular, benefited from an increasing globalisation of capital that created opportunities to operate worldwide in some of the most dangerous areas of the world. Along with these opportunities also went security risks, while to manage these risks, multinational corporations now started to look around for security companies with the expertise to provide solutions for the new types of security problems they now faced.

The rise of international terrorism during the same period also created opportunities for private military security. Such terrorist groups included:

> the Angry Brigade in Britain, The IRA in Ireland, the Red Army Faction in Germany, Action Direct in France, the Red Brigade in Italy, ETA in Spain, the Red Army in Japan, the Tupamaros in Uruguay, the Grey Wolves in Turkey, the Tamils in Sri Lanka and the various Palestinian factions scattered around the Middle East.[101]

In 1970 alone the US State Department recorded over 300 terrorist acts worldwide, while such attacks increased in numbers over the rest of the decade. It is estimated that 40,000 lives were lost to terrorist acts between 1970 and 1985.[102] Such attacks were not always targeted at the commercial sector. Even so, corporations could not afford to sit back and do nothing in the present climate. The threat of kidnap was particularly apparent, especially after the kidnapping and subsequent murder of the powerful West German industrialist Hanns-Martin Schleyer who was found shot dead in the boot of his car after Palestinian terrorists had failed to gain the release of members of the Baader–Meinhof gang, a group of middle-class revolutionaries then in prison in Germany.[103]

Even though Watchguard had already identified commercial opportunities in international security, it was Kroll, Control Risks Group, and later Defence Systems Ltd and Saladin, which were to benefit from the shift into the commercial market for private military security. As Westbury explains,

> Kroll supplied commercial investigation services for the corporate sector, Control Risks Group undertook political and security risk analysis for the same customers, Saladin Security concentrated on close protection for executives and Defence Systems Ltd focused on providing security services to supra-national agencies and multinational corporations.[104]

At the same time, such services provided solutions to complex security problems faced by their customers operating in dangerous regions of the world and confronting the threat of international terrorism.

By the middle of the 1970s these newly formed private military/security companies[105] were transforming the role of private military security through taking on commercial security operations. More importantly, they started to distance themselves from the clandestine operations that had been the hallmark of British mercenaries during the 1960s and into the 1970s. Mercenaries still operated, but these new companies stayed away from such operations instead concentrating on providing security solutions to the commercial sector. They still occasionally worked for the British government, taking on contracts that involved training foreign militaries, but such contracts were kept quiet because of the opposition they might attract from the press and general public in the UK. The shift into the commercial sector, which started in the mid-1970s and carried on throughout the 1980s and 1990s, is set to continue for the foreseeable future. Today, however, after establishing itself within the corporate world, the industry is now set to expand into the area of Security Sector Reform (SSR). No one should be surprised by this move. In many respects the shift is a natural one for these companies to make when considering the experience and skills in security they have to offer the sector. Moreover, since multinational corporations are already moving into this area, what is there to stop PMCs and PSCs from making the same move?

After the Cold War: new wars and strategic complexes in a new development–security arena

According to Kaldor, since the 1990s a new type of organised violence has developed, especially in Africa. Kaldor describes this new type of violence as 'new war'.[106] This type of war is very different to the type of war conducted between European states in the first half of the last century. Then, war was fought between states over the rights of a particular piece of territory. New wars are motivated by very different reasons. The goals of new wars are about identity politics in contrast to the geo-political or ideological goals of earlier wars.[107] What is meant by identity politics is a group's claim to power over another group on the basis of its identity.[108] The function of war is to exclude the members of the other group from society. On the other hand, some see new wars simply as a result of rational economic calculations. Here, violence is underpinned by economics. As Keen points out 'conflict can create war economies ... controlled by rebels or warlords and linked to international trading networks'.[109]

✶ How new wars are fought is also different to how war was conducted in the past. New wars are not fought on battlefields between opposing armies wearing uniforms. Instead, the battlefield is everywhere: cities, towns, and countryside. Neither is the behaviour of those who fight in new wars the same as the soldier who fights for a state military. The behaviour of a soldier is prescribed according to laws of war. Soldiers may not commit atrocities against non-combatants, destroy historical monuments, or act in an inhumane manner towards prisoners. In the case of new wars, there are no rules telling a combatant how they should behave. Thus, atrocities against non-combatants are normal in new wars. Such atrocities include mass killings and the systematic rape of women of all ages. It could even be argued that new wars represent a reversal of the processes through which modern states evolved.[110] Consequently, resolving the humanitarian crises brought on by new wars is beyond the capability of a single state.

For states to be able to respond effectively to such crises as a consequence of war they must now cooperate with a range of different actors. This cooperation has resulted in new crosscutting governance networks being established. These networks of strategic complexes involve state and non-state actors, from the local to the supranational level. Moreover, because many humanitarian crises are the result of new wars, alleviating the poverty of such crises is a much more difficult task. As Duffield explains, development discourse now recognises that poverty and conflict are interconnected, while reinforcing each other in different ways.[111]

A new development–security arena

The complex nature of development provision today has placed assistance beyond the capabilities of a single state. To be able to function effectively in the new humanitarian environment of new wars, cooperation between multitudes of organisations has been necessary. This transformation has been achieved through the establishment of new crosscutting governance networks involving state and non-state actors from the supranational to the local, and has involved the internationalisation of public policy. Furthermore, large areas of the developing world are now embroiled in conflict that has made relieving poverty very difficult. Indeed, according to Duffield, development discourse understands conflict and poverty as interconnecting in different but often reinforcing ways.[112] The association is usually presented in terms of poor countries representing an increased risk of conflict.

In many respects, the changing nature of development is as much to do with the determination of the industrial states to impose on the rest of the world a new liberal peace, based around the politics of humanitarianism. In liberal peace, the emphasis is on conflict resolution and prevention, strengthening civil and representative institutions, reconstructing social networks, promoting the rule of law, and security sector reform, while achieving all this within the context of a functioning market economy.

States remain important actors in development projects even though development assistance has become much more multipointed, bringing together governments, international agencies, and NGOs in new and complex ways. Consequently, there has been a noticeable move from the hierarchical, territorial and bureaucratic development to a more polyarchical, non-territorial, and networked approach.[113]

At the same time, because development assistance is occurring more frequently in countries threatened with, or actually experiencing, new wars, development has come explicitly to focus on security. Security in the context of development means human security, embracing vulnerabilities and risk to do with health, employment and welfare needs, while shadow wars have altered the development security complex. Shadow wars have turned development into a business, while creating opportunities for private security. The main role of private security is to support security sector reform that will bring to an end shadow wars, while also supporting multinational corporations and NGOs operating in areas affected by such wars.

Development and security converge

Contained within liberal peace and reflected in ideas on liberal war, is the blurring of development and security. The intended aims of liberal peace, together with willingness towards humanitarianism, symbolise this union, while the determination to resolve protracted conflicts and post-conflict reconstruction so as to prevent the possibility of war recurring characterises a significant change in developmental politics. Post-conflict reconstruction must ensure that the problems, which had previously ignited wars, do not do so again, as has happened with development in the past. For this to happen, society needs to change, while the process of change should not become matter of chance. Such far reaching changes to development are, at the same time, closely linked to the reproblematisation of security according to Duffield.[114] By reproblematising security in this way, the purpose of private violence is given meaning within the conventional views on the causes of the new wars, usually associated with a developmental malaise of poverty, resource competition and weak or predatory institutions. Thus, not only have we radicalised developmental politics, but, crucially, the position mirrors a new security framework within which we find private violence of the type that includes PMCs.

The encounter between the multifaceted natures of governance networks with the political dynamics of new wars has created a new development/security agenda. On the one hand, liberal values and institutions now have the power, or at least the professed goal, to improve the lives of millions of people living in poverty. Meanwhile, these powers are being deployed within a new security framework in which underdevelopment is seen as both destabilising and dangerous. Yet, even though this contested terrain, comprising complex relations of structural similarities, and, at the same time, new asymmetries of power and security, looks set to dominate our thinking on underdevelopment over the coming years, it is only now being taken on board by policy oriented development, while some areas remain out of bounds, notably the area of private security.

The privatisation of violence in the new development–security arena

Whereas before, violence was manipulated through the patron/client relationship, now strategic resources, which should be directed at development, are used to establish the necessary political and commercial networks around which violence has been privatised. Thus, strategic resources play an important role in strengthening the position of the ruler. Commercial concessions to multinational corporations, or local businesses, of resources such as diamonds, oil, and exotic timber are used to raise funds to buy support from strongmen and warlords. The revenue enjoyed from the sale of strategic resources is also directed at creating parallel military budgets that then pay for alternative security structures supplied by the private sector. For the public, however, the process by which violence is privatised implies the removal from the authority of the state of certain fundamental responsibilities towards whole swathes of society.

The changing nature of private violence in the developing state

One of the defining features of the nation state has been its successful claim over the legitimate use of violence within its borders.[115] The ability of the state to separate out public and private violence, and then criminalise private violence, has been overall very successful, especially in Europe. In developing countries, new centres of authority, as discussed earlier, have challenged the competence of the developing state. There is no longer a clear distinction between the law as a public service, its use for private gain, and its enforcement. Instead, these distinctions have become blurred and ambiguous. As a result, the effectiveness of existing laws, property rights, and customary practices has been eroded, if not totally undermined. This situation is not restricted to any one part of the community, but affects all social groups within a society. The rich, for example, find they cannot call on the bureaucracies of the state to protect their wealth, while customary law that should, for example, safeguard the rights of access to common land for the less well off, is conveniently forgotten in the interests of ruling elites, who hope to take advantage of such land through foreign firms. An inability to enforce the law also affects multinational corporations who have either to pull their employees out or risk their lives if they carry on operating inside countries where the government is unable to exert adequate control over internal security. Alternatively, they can themselves use private violence to protect their operations, as has happened in the past. In many respects, the policy aims of development and security and the business sector are rapidly converging. Both major international extraction companies and those international organisations responsible for development and security have a maximum stake in long-term stability, as well as having a high degree of exposure to public pressure.[116]

Thus, for many social groups in the developing world, private security has taken over as the primary source of protection. For those that can afford it, these groups must now rely on private protection to safeguard themselves and their properties. Without it, they become easy targets for extortion or even murder. At the same time, the tensions in these relationships have increased demand for all types of private security arrangements at the international level.

Turning development–security into business

An important characteristic of the new development–security arena has been the role allotted to the business sector. In terms of actual response, it is too early to say whether the impact of commercial companies on developing states has improved their position. What engagement there has been has generally come from multinational corporations, especially companies representing the extraction industry. By participating as full members of the governance network, these companies hope to enhance their business image and improve on their market position, while securing their business environment. The same can be said of the private security industry, especially PMCs determined to legitimate their activities. Moreover, by turning development into business the commercial sector has become a major player in strategic complexes.

Strategic complexes

Strategic complexes are made up of a range of different actors. These actors include governments, international organisations, charities, military establishments, PMCs, PSCs and the business sector. The purpose of strategic complexes is to promote global peace by providing stability and security in an increasingly hostile world. To achieve this, strategic complexes follow a radical programme of social change. Instead of being concerned solely with development the programme is also concerned with security, while the programme has privatised and militarised the activities of the participants.

The organisational form of strategic complexes is very similar to the organisational form of new wars. Increasingly, actors from both groups are from the private and not public sector and working in areas not associated with government competence. One of these areas is mine clearance which is being undertaken more and more by charities such as Mines Advisory Group (MAG) and PSCs such as ArmorGroup.

More importantly, strategic complexes have opened up a space for private military force. The security skills of PMCs and PSCs are rapidly becoming an essential component of strategic complexes. Needless to say, PMCs are at the forefront of any move by other actors to utilise such skills. This is presumably because of their military proficiency. Many former soldiers employed by PMCs have served in first world armies, particularly the American, British and French armies considered to be the finest armies in the world. At the same time the UK has an historical association with mercenaries who have a tradition of working for state and non-state actors. Finally, in respect to the UK as events in Iraq, Kosovo, and Sierra Leone have shown, the British government tends towards an interventionist stance, while this is reflected in the actions of some UK PMCs. For example, three UK-based companies Rapport Research & Analysis, Gurkha Security Guards, and Defence Systems Limited explored opportunities for providing security in Sierra Leone.[117] Leaders of weak states have used PMCs to retain control of strategic resources that, in turn, enables them to marginalise threats to their rule by denying opposing strongmen revenue to purchase weapons.[118] At the same time, international organisations, intergovernmental agencies, and multinational corporations have come to rely on the security skills displayed by PMCs to be able to operate in new wars, while also protecting their employees and assets.

Another feature of strategic complexes is their reliance on the commercial sector. In this respect, development and security is turning into a business. Private violence in the shape of private military security protects strategic resources, while the impact of private violence goes beyond just guarding a country's natural wealth. Such violence affects all social and economic groups struggling to survive the environment of new wars. Multinational corporations, especially those in the extraction industry, employ PMCs and PSCs to protect their employees and assets that can run into millions of dollars. Charities also use private violence to protect their staff and assets, though they tend to use local private violence, rather than the type of private violence supplied by Western PMCs. Even government agencies whose staff work in dangerous areas may rely on private violence. UK-based Control Risks Group and ArmorGroup escort UK FCO teams in Iraq, while DynCorp did the same for Paul Bremer.[119] Private security has taken over as the primary source of protection in many developing states where new wars are common.

Such a trend is the result of turning development security into business. Furthermore, the international community is likely to see this trend increase as more roles are allotted to the business sector. In Iraq, for example, US authorities have already outsourced security tasks to the private sector.[120] In future, they may require companies bidding on reconstruction contracts to include security in their proposals. Contractors would be required to make robust security plans ahead of time, while such a policy, if enacted, would provide a financial bonanza to private military/security companies such as MPRI, DynCorp, Oliver, ArmorGroup, Erinys International, Blackwater, Control Risks Group, Kroll, and Aegis. Taking over responsibilities for security is not restricted to Iraq, but the country is a good example of how governments in the West are increasingly coming to rely on private military security to achieve their objectives.

It is too early to say how the impact of business will shape the development–security arena. Neither is it possible to say whether the impact will be positive in every case. At present, what engagement has occurred has come from the construction and extraction industries. Each industry is becoming more reliant on private military security to protect their projects. These groups are now participating in strategic complexes, while the first two groups are increasingly being recognised as full members. The business image of construction and extraction companies working in this area is being enhanced and their market position improving. The same is starting to happen to companies selling private military security. But, improving the image of PMCs will take longer because of their historical links to the mercenary world, as a result of which full membership to strategic complexes is still not an actuality. How long this situation will continue is unclear.

Conclusion

This chapter has shown that private violence has existed for centuries. On the other hand, the nature of private violence has changed over time. What determines these changes are social forces, which all societies experience. The French Revolution released a new set of social forces that eventually led to the abolition of the practice of governments hiring foreign soldiers instead of conscripting their own civilians into the army. This in turn gave rise to national armies made up solely of citizen soldiers. During the Cold War, occasional examples of private military force still emerged; not least in the UK. An example of such private military force was the mercenaries used in the Yemeni civil war, the role of mercenaries in the Hilton Assignment, and the mercenary debacle in Angola. The end of the Cold War means countries are once again experiencing changes to their security arrangements. This time, however, private violence is increasing as public violence struggles to cope with the challenges it now faces, while such a change is most evident in Africa.

A new development–security arena is emerging in many parts of the developing world. Development and security can no longer be understood as separate issues but have converged into one, while violence is being privatised within this new arena. Such an approach, turning development into business, is the result of the inability of Western governments to confront this problem themselves. As a consequence of this, governments are entering into partnerships, known as strategic complexes, with other interested parties to confront the humanitarian crises occurring in this new arena and which are normally the result of new wars. This has opened up possibilities for PMCs and PSCs to provide security for the parties involved in strategic complexes if they are to achieve success.

3

THE ROLE OF PMCs
AFTER THE COLD WAR

Introduction

This chapter illustrates the nature and evolution of the PMC market after the Cold War. From the middle of the 1980s it was becoming increasingly clear to those working in the security industry that a new type of private security actor was starting to emerge on the international stage. The nature of this new actor was different to previous examples of private violence, even though in respect to the UK a clear historical link exists between them and the classic mercenary. At the same time the British government has been very slow to recognise this new actor, even with the publication of the Green Paper in 2002 proposing various options for regulating their activities.[1]

The chapter first discusses the impact of private security on British foreign policy after the Cold War. Initially, the government paid no real attention to the impact PMCs were starting to have on international security. This changed when the South African PMC EO undertook military operations in Angola and Sierra Leone. These operations allowed the international community to marginalise its responsibility to intervene in the civil wars going on in these countries. This was particularly so in the case of the British government's responsibility towards Sierra Leone. EO's inclusion in the chapter is also important because the company's actions have had a direct impact on the industry in general. Many of the ideas now being put forward by politicians, including the suggestion that PMCs can make a positive contribution to international peace and security, are a direct result of EO's involvement in Angola and Sierra Leone. Not to include EO would leave a large part of the PMC story untold, while other parts of the story would simply not make sense.

Next, the chapter examines the shift by PMCs into the commercial, as opposed to governmental, arena. This shift is further recognition of the newness of PMCs. Unfortunately, however, the shift exposed tensions between the British government's efforts to continue to control PMCs through unofficial channels and the companies' own commercial agenda that no longer permitted them to stay in the shadows of a secret world. Indeed, as the 'arms to Africa' affair demonstrated, the shift to the commercial arena finally saw a rift occur between the secret world and liberal Britain, which for so many years, the government was able to keep separate from each other.

The chapter then briefly examines the transformation of the legal personality of PMCs, and what this means for the future of the industry in international security. The legal transformation of PMCs is not restricted to UK PMCs only. More importantly, with respect to MPRI, they are much further ahead in transforming themselves legally, a process started over 50 years ago, and that UK PMCs are now trying to emulate. EO, on the other hand, was also very successful in constructing a legal personality and because of this its methods are frequently copied by other companies entering the industry. The final part of the chapter examines the British government's failure to recognise that the nature of its relationship to UK PMCs had changed by the 1990s. As a consequence of this, even though UK PMCs were by then having an increasingly positive impact on international security, signing contracts with governments in developing countries or with multinational corporations, the government still sought to distance itself from them. By doing so, the government added to the gravity of the 'arms to Africa' affair as explained in Chapter four.

Foreign policy and private security after the Cold War

The changes brought about by the dramatic events surrounding the end of the Cold War saw the need for governments to revise their foreign policy. Much of the world now faced a new set of challenges, many from the Third World. How, for example, was the international community to maintain order in a world no longer sure about how the new distribution of power was going to play out? For Britain, the challenges were reflected in the underlying principles of British foreign policy. There was still a clear understanding that Britain should be a major player on the world stage, that its commercial interests should be protected and that the rest of the world could not be ignored.[2] Meeting these challenges was not always going to be easy for a government constrained by international law, questions of ethics, domestic policy, the media revolution and costs. It is unclear whether the government did seek an alternative approach to circumvent these problems to facilitate tackling the challenges it now faced. What is clear is that the introduction of private security as an alternative to state military aid was not initially seen as a problem by governments, nor did it undermine the underlying principles of Britain's foreign policy. This was the situation until Sandline International's involvement in Sierra Leone. Then, deep-seated tensions emerged over the actual nature of the company as a result of the confusion over the UN arms embargo.

One of the biggest problems that faced not just Britain but the whole of the international community was to decide how to respond to the civil wars now emerging in different regions. There was already occurring at this time a shift in the attitude of the international community towards cooperation between states. This was exemplified in the international community's broadly conceived attitude towards humanitarian intervention to protect basic human needs and rights. But, if these needs and rights were to be protected, ideas about responsibility and security had to be revised to take account of the problems derived from economic development, environmental management, human rights protection, and multinational peacekeeping, preventing ethnic conflict and with whom the responsibility for resolving these problems laid. The international community's initial optimism for cooperation was exemplified by the decisions to intervene in Northern Iraq in 1991, the Balkans in 1991 and Somalia in 1992. Beyond these three interventions though, the UK and US refrained from further participation in humanitarian operations that required substantial numbers of troops but that were not seen as essential to their own national interests. At the same time, they raised little objection to intervention by others, including non-state actors, especially in Africa, if it meant they were able to reduce or transfer their moral responsibility to someone else.

Participation in helping to resolve the civil wars going on in Africa throughout the 1990s was, in most cases, limited. The West's failure to respond adequately to the 1994 genocide in Rwanda clearly shows a general lack of concern over the problem of war in Africa during this period.[3] Only after the genocide, and in response to the massive flight of people from Rwanda, did the West become more generous with offers of help.[4] Even before this, the West's attitude to the crisis can be gleaned from Britain's Ambassador to the UN, Sir David Hannay, who opposed reinforcing the UN peacekeeping operation because of the Somalia debacle. Instead, the Ambassador suggested withdrawing the UN peacekeeping troops, leaving behind a small contingent to try to negotiate a ceasefire.[5] More important, the task of withdrawing from the crises that followed Rwanda was made even easier by the introduction of state sovereignty into the debate on the right to intervene in the affairs of other states.[6] They could rightfully argue that what went on inside another state had nothing to do with them. Britain, in particular, benefited from this. As a former colonial power and a permanent member of the UN Security Council, with the ability to project military force overseas, one would expect the country to accept limited responsibility to intervene to stop some of the fighting ravaging the African continent. Instead, they chose to do as little as possible. It is not surprising, therefore, that under these conditions African governments decided to employ private forces to fight on their behalf. After all, recourse to private forces was a logical avenue for them.

At the same time, the increased frequency of civil wars in Africa opened up a whole spectrum of commercial opportunities for the security industry. The industry had already become a growing and accepted feature of life in the West,[7] as Western governments continued to recognise the industry's changing behaviour. The industry's domestic agenda included guarding premises, protecting cash transfers from banks, and guarding important people. What was less known was their involvement in international security. Now, PMCs were able to exploit new opportunities as a result of the changes going on within the wider sphere of the international community. Companies started providing military training and assistance to developing countries in particular. For example, from 1993 to 1994 the Gurkha Security Group provided training to the Sierra Leone Army.[8] The company pulled out of the country after suffering losses, including its commander, in an RUF ambush. Other PMCs supplied military equipment or advice, were prepared to give military threat assessments and organised logistical support. Such was the demand for the military services PMCs could supply, brought on by the increasing number of civil wars in Africa, that increasing numbers started to appear on the international stage.

By the 1990s, there was emerging a clear commercial agenda for private military services. Indeed, the US government was keen to use US PMCs to promote its foreign policy in the Balkans because of the limited political risk. It therefore gave support to US companies in their bids for military contracts.[9] Such a move implied the US government was confident US PMCs would have a positive impact on the region's security situation. The British government, on the other hand, refrained from using PMCs in this way. Instead, they chose only to monitor UK PMC activities. This, in turn, allowed them to deny any direct involvement with the companies; a decision the government would come to regret.

The majority of customers were now either leaders of Third World countries facing political crisis or multinational corporations from the oil and mineral extraction industry. These groups were later joined by international charities. As the following discussion explains, politically weak leaders and multinational corporations established flexible business arrangements whereby, in return for drilling or mining concessions a political leader could afford to purchase the services of a PMC to protect the country's strategic resources and, at the same time, marginalise any threat to them by strongmen.[10] Again, this arrangement has benefited Western governments responsible for keeping open the flow of strategic resources to Western markets. Furthermore, it is not beyond one's imagination that to ensure this, Western intelligence agencies and PMCs worked together where there were shared interests at play. Charities, on the other hand, tended to rely on local security guards to protect their operations, distancing themselves from Western PMCs. The exception to this is humanitarian demining where Western PSCs have been employed to undertake demining operations. This topic is covered in greater detail in a later chapter.

The experience of EO

One of the first companies to exploit the commercial opportunities offered as a result of the changes that occurred to the political and security environment in the early 1990s was EO. The privately owned military company was founded in 1989 by Eben Barlow, a former assistant commander of the 32 Battalion of the South African Defence Force (SADF) before being employed by the South African Civil Cooperation Bureau (CCB).[11] As discussed in more detail in the section covering the legal transformation of the company, EO was initially set up to provide intelligence training for the SADF Special Forces, while most of its personnel, with the exception of some specialists such as Ukrainian pilots, tended to come from the same Special Forces background.[12]

The company undertook its first significant military operation in Angola in 1993, 'which led the charge of the new corporate brigades into Africa'.[13] They were first employed to capture and defend valuable oil tanks at Kefekwena and then to repeat the operation for the oil town of Soyo, which had been taken over by UNITA troops. However, UNITA troops recaptured the town when the company later withdrew. The operation, according to Shearer, was nevertheless significant in that it was the first real demonstration of EO's combat capability.[14] Their client, it appears, was Anthony Buckingham, a senior board advisor to several North American oil companies, suggesting that the operations were designed to look after the interests of multinational corporations Buckingham represented. Even so, the government of Angola also benefited from EO's successful operation.[15] This initial success also demonstrated how useful the company's military skills could be in helping to deter the increasing military threat posed by opposition forces. This, in turn, led to the Angolan government offering EO a one-year US$40m contract in September 1993.[16] The company was to train and supply weapons to government troops. The contract was renewed for 12 months in September 1994, and for a further three months in 1995.[17] Their contribution to governmental success on the battlefield temporarily improved the country's security since it prompted UNITA to sign a peace accord in Lusaka in November 1994.

Such success was not lost on Western countries either. From the very start of the company's operation in Angola, there was no attempt by any government, particularly in the West, to have EO withdrawn. Instead their presence allowed Western governments to conveniently marginalise Angola's troubles.[18] Even when President Clinton did intervene, suggesting that EO's presence lay behind Savimbi's refusal to adhere to the Lusaka agreement, it is more probable that this move was linked to a rival American company, MPRI, wanting to win contracts over EO.[19] There was, of course, no reason why an outside government should interfere in the country's affairs. After all, the company had been contracted by a sovereign state that could do more or less as it wished inside its own borders. EO was also providing protection for Western business interests in the shape of oil fields and diamond mines. What the company's operation did prove was that, under certain conditions, military operations could be made commercially viable. As such, EO could be a useful tool for countries in the West with the ability to project military power overseas but not always keen to do so because of the cost to the treasury and the political problems that invariably came with these types of operations.

EO's operation in Angola established it as an effective PMC able to supply security specialists to governments to help train their security forces and provide military protection against local insurgencies. It also proved a PMC could have a positive impact on a country's security situation. This is probably what influenced the Sierra Leone government to hire the company. In the case of Sierra Leone, the initial contract was signed in April 1995 with the National Provisional Ruling Council (NPRC). EO was to provide military support and advice to the Sierra Leone army, which was engaged in a civil war with the RUF. The company arrived in Sierra Leone in May 1995. By this time, the RUF was only 20 kilometres from the country's capital Freetown. The company was set four objectives:

- To secure Freetown
- To regain control of the country's diamond fields and Rutile mine, thereby allowing the government to generate revenue and help to guarantee EO payment
- To destroy the RUF's headquarters
- To clear remaining areas of RUF occupation.

Within eight months these objectives had more or less been achieved, forcing the RUF to agree to negotiate with the government for the first time in five years.[20] EO spent 21 months in Sierra Leone during which time the British government appeared to have shown no interest in the company's military operations. From 1990 to EO's official departure in January 1997, the British government took little interest in the country's military problems, even turning down a request by the local government for military support in May 1991.[21] Finally, it was not that Britain had no interests in its former colony, but by not objecting to the introduction of EO into the country, the British government was able quietly to abrogate its responsibility towards the people of Sierra Leone. Instead, EO took on the responsibility of securing a peace, while the local government carried the cost of the operation. On the other hand, Britain's commercial interests in the country's diamond trade were now safe as a consequence of EO's intervention on behalf of the local government, while a healthy distance was maintained between the British government and EO.

While EO's operation in Sierra Leone again drew a general lack of interest from national governments, other elements of the international community were starting to take notice of the increasing role of private violence in ongoing conflicts, especially in Africa. This was evident from UN Resolution 1995/5 of February 1995, the Commission on Human Rights, which reaffirmed that the recruitment, use, financing, and training of mercenaries, should be considered offences of grave concern to all states. This was reiterated in decision 1995/254 of 25 July 1995, when the Economic and Social Council approved the Commission's decision to extend for three years the mandate of the Special Rapporteur on mercenaries.[22] Even so, many states still felt that domestic problems were the responsibility of the country concerned and not necessarily that of the rest of the international community. Moreover, in the case of the UK policy on Sierra Leone, the government could avoid sending troops to the country while EO was there.

Yet this international reaction needs also to be seen in its political context. Had EO not been a South African based PMC, rooted in the apartheid era, it is possible they would have drawn minimal attention from the international community. Moreover, while leaders from developing countries condemned their actions, labelling them mercenaries and criminals, Western governments were more measured in their response, particularly the US and UK governments.[23] It is possible that EO's presence in Angola and Sierra Leone influenced Western governments to not get involved militarily in the country's civil war, though it is unlikely any government would acknowledge this fact. For the US and UK governments, however, EO's presence allowed them to marginalise or simply ignore the provision of military support.

The rise of the PMCs

By the mid-1980s, PMCs started to transform themselves into properly structured corporate entities. The permanent nature of these corporate entities was seen as a key shift away from earlier *ad hoc* mercenary organisations, and linked to their legal transformation. This area is discussed in greater detail in the next section covering the legal transformation of PMCs. The biggest change, however, did not occur until after the end of the Cold War. It was then that the term PMC came into use, marginalising the word mercenary. Throughout the 1990s, PMCs gradually established themselves as commercial enterprises separate from any involvement with governments, except when their interests converged. Business profit was now the essential driving force behind PMCs.

Furthermore, the drive for business profit as opposed to individual profit emphasised the newness of PMCs. As Spicer explained in 1999 as the Chief Executive of Sandline International, 'we are a commercial organisation that exists to make money, with a number of backers and a number of partners'.[24] As such, the company, along with numerous other military and security companies at this time, sought to distance themselves from the mercenary label if they were to be successful. But to have any chance of making a positive contribution to international peace and security the companies had to ensure their newness was recognised by the international community. To do this, they had to transform themselves into legal entities operating openly.

The social, political and economic changes going on in the developing world increased the need for more PMCs; a process which continues to this day. In particular, the changing nature of state competence, the growth of the private sector, market deregulation and military downsizing all contributed to the rise of PMCs on the international stage.[25] This last factor saw large numbers of soldiers released from service with the necessary skills to fill the gap for the demand for private protection. Other security companies benefiting from these changes included DSL, Control Risks Group, and Rapport Research and Analysis. The first two companies had in fact been established in the 1970s–80s. With the end of the Cold War, business opportunities for them expanded. However, according to Shearer, 'these companies were primarily concerned with providing security services, and not direct military assistance as in the case of EO or Sandline International'.[26] As we see below, this type of security service lent itself to protecting economic assets, vital to the continued prosperity of the country. A number of these security companies also undertook to support British foreign policy, as they had done in the past, by taking on contracts that were seen by the government as politically too sensitive. Indeed, both were able to benefit from the other. As an experienced British security consultant noted in July 1997:

> The Foreign Office has a list of companies that are competent in carrying out training to whatever standard, whether it is counter-terrorist work or just general military training. If a request for British military training is viewed as politically sensitive, the Foreign Office will say, 'these companies can handle it'.[27]

It is within the economic sphere, though, that these security companies appear to have had the most influence since the end of the Cold War. As Carr-Smith of DSL explains:

> Our clients include petrochemical companies, mining or mineral extraction companies, multinationals, banks, and embassies ... . Very often the sort of first-in-type companies that are trying to get things going. We provide them with a service that allows them to operate wherever they are.[28]

The position of government agencies to protect Britain's economic interests was explained by The Lord Chancellor, Lord Mackay, when he stated that:

> MI6 is involved in protecting the economic wellbeing of the country by keeping a particular eye on Britain's access to key commodities, like oil and metals, while useful commercial intelligence collected and collated by MI6 is passed on to Britain's major companies.[29]

It is clearly evident from both these statements that they express a common interest. The former is concerned with the physical protection of the assets of British companies involved in the oil and mineral extraction industry; while Lord Mackay sees the continued access for British companies to key commodities, including oil and mineral resources, to be a priority for the intelligence services. In many respects, both groups are opposite sides in a mutually supporting relationship. We should not be surprised at this when we consider the following statement by a former SAS officer now employed by a private company, who noted in mid-1997 that he worked much as he had before, but 'on the other side of the public–private fence'.[30] The statement is not intended to suggest that during the majority of the 1990s security companies were complicit in the covert affairs of the government, but that a close working relationship can exist between both groups where they share a common set of interests, as with the protection of the country's economic interests.

The transformation of the legal personality of PMCs

By now, there were already changes occurring in the industry that sought to distinguish PMCs from mercenaries; recognising the legitimate right of the former to exist while criminalising the latter. Companies were by the 1990s operating openly for major clients, including governments and multinational corporations. The most important of these changes was the transformation of the legal personalities of mercenaries, which had started two decades previously with Watchguard. Then, the mercenary represented a figure similar to other criminal individuals, in that they represented the projection of illegal force, a contract killer, while today the security advisor working for a PMC only undertakes legally constituted work from clients. Both these descriptions represent opposites on a continuum, with the middle still under discussion as to the legal status of the companies that occupy this area.

By the 1970s, British companies in particular were entering into legally binding contracts to supply military advice and training. At the same time, they were still prepared to recruit classic mercenaries since they were easily deniable if a contract went wrong.[31] Even so, it is doubtful whether this combination of a legally constituted company employing mercenaries on short-term contracts was ever going to work, since customers increasingly had to ensure that any individual who worked for them was a legally contractible party. A mercenary by definition does not fall into this group of a legally contractible party. Using the term security advisor would help security companies put distance between them and mercenaries. But in the long term, much more was needed to demonstrate to customers that these two actors belonged to separate groups.

By the 1980s, the companies were themselves describing their employees not as mercenaries but security advisors, while the type of short-term contracts based on cash payments, which Stirling may have offered his French mercenaries, were no longer suitable for PMCs and some PSCs. Furthermore, with greater financial scrutiny by government departments, cash transactions, which mercenaries prefer since they are guaranteed payment, were not a suitable way to conduct business. Instead, companies would pay the same way they paid all their employees or as specified in the contract. Each method is open to inspection by government officials, notably the taxman, again creating the need for a legal relationship between employer and employee or contractors. Thus, by the late 1990s, a security advisor working for DSL was very different from a mercenary working for Watchguard.[32] This transformation is ongoing.

There are a number of reasons for this shift. First, if they were to prosper financially, it was necessary for these legitimate businesses to distance themselves from the mercenary label associated with the Angolan debacle, which was still fresh in people's minds. Second, as business grew it was necessary to conduct it within legal boundaries at all levels, including in the field. As such, employees could no longer be referred to as mercenaries, since to do so would imply the law having been broken in connection with a legitimate business operation. For this to occur, employers had to take greater responsibility for the employees, including those working in the field.

During the late 1980s, companies started to adopt a legal personality that would distinguish them from former mercenary armies. EO was one of the most successful of these companies and although it was not a UK company *per se*, it bears examination for the transformative role it played and because other companies started to copy the way it constructed its legal personality.

> as a privately owned military group whose finances, personnel, offensive operations, air wing division, and logistics are all handled within a single group or through interlinked companies and enterprises, [a]nd in its most basic form, it would be managed by a common pool of directors and have a small permanent corps of staff, serving its own commercial interests and those of affiliated entities. Such a group of companies would typically be owned, organised, paid and deployed by the controlling shareholders

of one or more private companies which, in turn, may be transnational conglomerates.[33]

The company[34] was originally a front company providing intelligence training for the SADF Special Forces.[35] With its first significant operation in Angola in 1993 being a success, however, the company soon became a serious business concern attracting the interests of other parties. According to Pech:[36]

'[it came] under the primary control of a group of British-based entrepreneurs, ... [a]nd a consortium of military-related and mineral companies was established. EO formed a key military component in this group ... . EO grew into a veritable military giant between 1994 and 1997 whose operations in Africa were facilitated by these affiliated companies.[37]

It is clear that EO's structure was very different to the mercenary organisations of the 1960s and 1970s. Instead of representing an *ad hoc* structure, the company was part of a consortium, the business organisation of which closely resembled a loosely connected network. This arrangement allows companies from the same consortium to come together for the discharge of specific tasks. The network makes available an array of expertise and capabilities that are important in achieving a successful outcome. Thus the organic nature of the structure allowed EO to respond to rapidly changing conditions on the ground.

The argument that PMCs have legal personalities and that this was and still is a key feature distinguishing them from mercenaries is in part reflected in the type of person PMCs employ. Spicer understood the significance of this the moment he helped create Sandline International.[38] In particular, he emphasised the permanent nature of the company and its legal personality. As he points out:

a private military company [including his own] is a permanent structure with a large number of people on its books. It has a permanent presence, it has an office, it uses promotional literature, it has a vetting system, it has a doctrine and it has a training capacity, internally as well as externally. It draws on the normal support that you would expect from a business; it is the official military transformed into a private sector in a business guise.[39]

The first three distinctions certainly point to a legal personality, while the remainder enhances that personality. If we now look at Spicer's definition of a mercenary, we again see a clear distinction between Sandline International as a company and mercenaries operating within an *ad hoc* structure. As he explains:

mercenaries are usually individuals, recruited for a specific task. They have no permanent structure, no group cohesion, no doctrine, and no vetting procedure. Their standards, both behavioural and technical, are somewhat suspect and their motives can be questionable.[40]

While some of these distinctions can be questioned – for example, mercenaries tend to only work with people they know thus ensuring some type of vetting procedure[41] – they highlight the point that mercenaries do not appear to represent a legal personality in the same way a commercial security company does. At the same time, while we can certainly describe a company such as Sandline International as a legally constituted company, the question remains as to whether their employees engage in mercenary activities. This particular subject is discussed in Chapter 4 the 'arms to Africa' affair, and the subsequent reports that followed the investigation.

While for many the distinction between Sandline International and mercenaries is confusing, rarely it appears does such confusion occur in the case of MPRI. More interestingly, even though the distinction is a global trend, it is not an even trend and some PMCs are further down the transformative path than others in different countries. In the case of MPRI the company has a corporate structure with a very easily identified legal personality. The company was established in 1989 with its corporate headquarters in Alexandria, Virginia, US. By 1998, the company had more than 400 employees, with a turnover of US$48 million for 1997. It is able to access more than 7,000 retired soldiers and officers ranging from general officers to platoon commanders, while its activities have included teaching basic training to soldiers, to the production of military doctrine.[42] In this respect, the company defines its mission and vision as follows:

> MPRI's mission is to gather and employ the experiences and talent of the top of the national resources of former military professionals, primarily in defence related areas, for the US government, and to assist foreign governments in converting their military into Western models that support democratic institutions … . The MPRI vision is to create a company of high quality professionals, tapping the cream of former/retired military, that sets the standard for quality performance, contract flexibility, business ethics, and management integrity, a company that makes a positive difference for those with whom it works, a company that grows in revenue and reputation; a company with worldwide recognition and respect. [43]

Again, adopting a legal approach to business does not necessarily mean that the company has not allowed its field operators to engage in mercenary activities in the past or even today. In the case of MPRI, however, the chances of mercenary activities happening have been reduced by a further set of reasons. First, the majority of work the company has undertaken has been in the domestic market and all of its contracts are undertaken overtly;[44] virtually all mercenary activity is conducted covertly, distinguishing MPRI from the mercenary. Second, the company has not involved itself in combat operations to date, only undertaking military training and consultancy contracts, drawing a clear distinction between itself and EO in particular.[45] Third, the company has never provided armed personnel, emphasising instead the company's policy that its employees should remain unarmed.[46] This last point is particularly interesting. The reason Soyster gives for prohibiting the carrying of weapons is that not only is it a bad business practice but there is also the problem of managing liability.[47] The suggestion of possible litigation has meant the company has had to take its legal relationship with all its employees very seriously. This is not to suggest other companies do not but the possibility of facing legal challenges has to some degree stopped MPRI from conducting combat operations, which by their very nature include arming people. As a result of this, a distinction exists between MPRI and companies such as EO or more recently Sandline International, one of which has conducted offensive operations. More importantly, refusing to arm their personnel sets them apart from previous examples of private violence. Even so, some individuals still see the work undertaken by MPRI as mercenary in nature.[48]

All the companies examined above have a legal personality. In this respect, they are very different to the individual mercenaries of the 1960s/1970s who might receive a contract from a leader in the developing world and then arrange a group of mercenaries to fulfil the contract. Whether it is possible to describe their work in terms of mercenary activities is much harder to answer. Even companies that have sought to conduct their relationship with their employees along proper legal lines, while prohibiting them from using weapons in all but the extreme cases, are still open to this charge.[49] More so when considering PMCs operate on an *ad hoc* basis, using databases to access former officers and soldiers, rather than having a permanent employee register. In this sense, they are still a legacy of past mercenary companies. What we can say with some clarity is that the type of private violence talked about here has been able to mutate and reinvent itself over the last 40 years. This has been necessary for private violence to be able to fit in line with the changes that have gone on within the social/political and legal environment or, as Pech reminds us, when private violence encounters challenging or hostile influences.[50] The changing nature of violence is of course not new, nor restricted to the private sector. To describe today's armies as identical to the armies that fought in the trenches in the First World War is ridiculous. Even the social value placed on violence has changed. Whether the activities of the above companies or the many other security organisations are mercenary in nature no longer appears relevant if the type of work they undertake is perceived as legitimate. How this position has been achieved is examined in the following chapter, the focus of which is the two investigations into the 'arms to Africa' affair, as well as the subsequent debates in the House of Commons that resulted in the publication of the Green Paper.

UK government confusion grows over the nature of its relationship to UK PMCs

The structure of the relationship between PMCs and the British government that emerged in the 1980s and into the 1990s was informal, in that there were never any written rules or guidelines established as to how the two sides should conduct business with each other. During the 1960s and 1970s this fitted nicely with the need to keep the world of secrecy apart from liberal Britain. By the 1980s and

1990s the interests of the companies were converging with the interests of the government. It seemed only natural, therefore, for them to cooperate so as to boost the interests of the country as a whole. At the same time, where a company's interests diverged, as long as those interests did not undermine the interests of the country there is no evidence to suggest a company felt obliged to inform the government of what contracts it was entering into. This was the position taken by Sandline International over its involvement in Papua New Guinea in 1996.[51] This marked a considerable shift in the relationship between the two groups from Stirling's days in Yemen. Then, there existed between government officials and mercenaries a homogenous relationship that was very different to the type of relationship that now exists between the government and UK PMCs.

The confusion over what PMCs represented now came to the fore. While the companies sought an open relationship with the commercial world, leaving behind the secret world, government officials had not realised that such a shift had occurred. While the Legg and Ibbs 'Report of the Sierra Leone Arms Investigation'[52] only describes the company as a PMC operating in the field of military and security service,[53] in the Foreign Affairs Committee report on Sierra Leone, the committee felt it necessary to describe Sandline International initially as 'in essence a company of mercenaries'[54] even though the company was trying to distance itself from the mercenary label. What followed next was a lengthy investigation into its affairs in Sierra Leone to prove the company had acted within the law. The outcome of the report cleared the company of any wrongdoing. The consequence of this was to legitimate further the company and other similar companies, while distancing them from the mercenary label and diminishing their need to remain secret, which had dominated the business for so long.

Conclusion

By the 1980s, PMCs had started to transform their legal personality to take advantage of the commercial opportunities that were now starting to appear in many parts of the developing world. By the early 1990s, they were starting to operate openly for government agencies and multinational corporations, undertaking a range of security related tasks from military training to supplying military equipment. Such tasks filled the security vacuum left by the downsizing of state militaries after the end of the Cold War. At the same time, PMCs were contributing to the security arrangements of some developing states. In the case of the Angolan and Sierra Leone civil wars, EO's success on the part of the government forced the rebels to the negotiating table.[55]

However, there still existed the problem of legitimacy, which had to be tackled if PMCs were to prosper financially in this new commercial environment. Unfortunately, with security companies already moving into the commercial sector and working for government agencies and multinational corporations the problem of legitimacy appeared to lose importance. Why Western governments ignored the problem of legitimacy could be attributed to a lack of interest or an inability to oversee the increasing number of security companies now emerging.

Whatever the reason, neither triggered a government response. Moreover, in the case of the UK, the move by PMCs into the commercial arena exposed tensions between the government's effort to continue to control PMCs through unofficial channels and the commercial agenda of companies that no longer permitted them to stay in the shadows of a secret world. Such tensions continue to influence the nature of the relationship between these two groups.

Indeed, during the first part of the 1990s, the relationship between PMCs and governments was changing. This was identified elsewhere, especially in the US with MPRI, but the UK government was slow to react to this changing relationship. Even though the Green Paper later attempted to deal with the issues raised, they have yet to be satisfactorily resolved in the UK. Furthermore, by initially ignoring the problem of legitimacy, the government only added to the gravity of the 'arms to Africa' affair in 1997.

4

TENSIONS EXPOSED IN THE 'ARMS TO AFRICA' AFFAIR

Introduction

The 'arms to Africa' affair was a sequence of events in 1997 that saw the exiled President Tejan Kabbah of Sierra Leone turn to Sandline International for help in returning him to power, and which helped to create the conditions that led to a turning point in legitimacy for PMCs. Immediately after the event a very gradual, but subtle, change started to occur in government thinking on whether PMCs should support British foreign policy. This change is still ongoing; while there is still unease among government officials about the role of PMCs in international security. That said, the question addressed throughout the chapter is how Parliament moved from a position of a general dislike of mercenaries, expressed in numerous statements made by members of parliament (MPs) during House of Commons debates,[1] to arrive at a position of one of general acceptance of the idea that PMCs[2] may have a function to play in international security as outlined in the Green Paper.[3] As the chapter explains, PMCs have sought to achieve a status similar to other forms of legitimate military enterprise, such as arms production, prominent in the UK, while MPs argue that their activities should be regulated in the same way as companies responsible for the export of arms.[4] In this respect, the Green Paper simply acted as a catalyst for change, though the nature of that change is still not fully understood.

To explain how this shift came about, the chapter will focus on a number of important issues covered throughout the different sections. The first section discusses the Sierra Leone Civil War, while the second section is a detailed examination of the role of Sandline International in Sierra Leone's attempt to defeat the RUF. The focus of the second section is first on the failure of Economic Community of West African States Monitoring Group (ECOMOG)[5] to defeat the RUF by December 1997, when Tim Spicer of Sandline International and Kabbah met, before discussing the UN arms embargo and Order in Council. Here, two key themes in particular are examined: first the ambiguity over which group the UN Resolution applied to, then the cause of the confusion over the promulgation of the Order in Council. An account is then given of the role played by Sandline International in supplying weapons to the conflict with the intention of using the weapons to defeat the RUF. The final part of this section examines the investigation into the affair. Following on from this, the chapter looks at the root cause of the confusion over Sandline International's identity, before bringing the empirical and thematic discussion together. The remaining sections explain the transformation of legality of the activities undertaken by PMCs that went on in the early 1990s, and how this, in turn, shifted the status of PMCs to the status of business organisations undertaking legitimate business activities. The final section also discusses the significance of the 'arms to Africa' affair on the industry in general.

An account of Sierra Leone's Civil War[6]

The civil war in Sierra Leone began in 1991, when the RUF, aided by Liberia, launched an attack on the then President Joseph Momoh. An appeal for military aid was turned down by the British government, leaving the Sierra Leone Army to hastily recruit new troops, many of whom were young and untrained. Without adequate supplies, training, or logistical support, these new troops were unable to defend effectively parts of the country from incursion by the RUF. In August of the same year, the Sierra Leonean people voted overwhelmingly to introduce a multiparty system, which led to the 1991 Multiparty Constitution being endorsed and passed into law, but this had little impact on the RUF, which continued its military campaign against the government. In April of the following year, unhappy with rebel operations in the diamond-rich areas in the southeast that indicated a new strategy of attacking strategic economic targets, junior officers staged a bloodless coup. There now began a period of unsuccessful military rule under Captain Valentine Strasser, followed by a brief spell under Maada Bio, who ousted Strasser in order to ensure elections occurred, which lasted until February 1996 when presidential and legislative elections were held.

The events that led to Sandline International's involvement in Sierra Leone started with the promise of elections in January 1996. These elections were, in part, made possible because of the successful military operations that EO had launched against the RUF commencing in early June 1995, as a result of which the RUF was eventually forced to the negotiating table. In the presidential and legislative elections, no presidential candidate received the required 55 per cent vote necessary for an outright win. In a run off, Ahmad Tejan Kabbah of the Sierra Leone People's Party (SLPP) was declared the winner with 59.9 per cent of the vote.

Over the following 14 months, the Kabbah government attempted to establish a peace agreement with the RUF. At the same time, Nigerian and EO operations increased the pressure on RUF forces to enter into negotiations with the government. Then, in November 1996, the government agreed to grant a general amnesty to RUF fighters in the context of the Abidjan peace agreement, which was finally signed on 30 November. The country continued to experience political unrest, even though one of the preconditions of the Accord, inserted at Sankoh's request and backed by international pressure, was the removal of all combatants and the withdrawal of EO. EO was served notice to terminate its operations in accordance with its contract in December 1996 and departed at the end of January 1997,

73

but not before warning the President that their total withdrawal, without leaving behind a small policing or intelligence team, would expose the government to a potential coup. During the last couple of months of EO's presence in the country the company's intelligence operations uncovered two attempted coup plots. Then, on 25 May 1997, another group of junior officers staged a coup and joined forces with the Armed Force Revolutionary Council (AFRC), seizing power from President Kabbah, who fled to Guinea.

In August 1997, the leaders of ECOWAS imposed sanctions on petroleum products, arms imports, and international travel of the AFRC, while the UN accredited President Kabbah as leader of the Sierra Leone delegation to the UN General Assembly. By the beginning of October that year, the UN, through Security Council Resolution 1132, had also adopted sanctions on weapons and other military equipment, petroleum, and petroleum products to the country. Later that month, in negotiations in Conakry between the Junta and ECOWAS C-5, and which resulted in the Conakry Accord, the AFRC/RUF agreed to restore Kabbah to office within six months, that is, by 22 April 1998, and to disarm. In November, the Nigerian foreign minister, Ikimi, on behalf of ECOWAS, requested the Security Council provide military and technical assistance to ECOMOG.

It was at this point that Sandline International's involvement began in earnest. In December 1997 Sandline's representative Tim Spicer met with Kabbah, who by now recognised the constraints placed on ECOMOG, in particular the lack of an explicit UN mandate and a crucial lack of resources for logistics and armaments.[7] It was then that Spicer proposed a military plan to help restore civilian rule to the country, which became known as the 'arms to Africa' affair. Spicer believed the company was well placed to give Kabbah's government the military help it needed to eject the junta. Kabbah knew EO, which sometimes collaborated with Sandline International on business opportunities in order to maximise the resource and expertise potential offered to a client.[8] EO had performed well in the 21 months it had spent fighting the RUF. Sandline International also possessed information about personalities, events, and trends in Sierra Leone that, if necessary, could be used to good effect, while the information may have been passed on to the FCO at the same time.[9]

The arms embargo

At the time of the coup in May 1997, the British government's policy was to give all necessary support to the restoration of President Kabbah's democratically elected government by peaceful means. The British government regarded President Kabbah's government as the legitimate government of Sierra Leone, even while in exile.[10] This position was reinforced when, in a reply to a personal plea by President Kabbah, the Prime Minister Tony Blair refused to sanction the use of force as the most appropriate response.[11]

At the same time, ECOWAS announced a subtly different policy, which was a total and general embargo on the supplies of oil and arms to Sierra Leone. However, as the Legg Report explains, the coverage of this embargo was qualified by an exception, which provided that the embargo imposed by this decision shall not apply to arms, military equipment and military assistance for the exclusive use of the sub-regional forces which shall be responsible for applying the measures contained in the Final Communiqué of the meeting of ECOWAS Ministers ... on the 26 June 1997.[12]

ECOWAS decided on a three-pronged approach that included diplomatic negotiations, economic sanctions, and the use of force. Meanwhile, the UN worked on a draft resolution intended to help remove the military junta. The Security Council finally passed the Resolution, No. 1132, on 8 October 1997. The Resolution gave additional support to ECOWAS. For example, the third paragraph reads:

[The Security Council] expresses its strong support for the efforts of the ECOWAS Committee to resolve the crisis in Sierra Leone and encourages it to continue to work for the peaceful restoration of the constitutional order, including through the resumption of negotiations.[13]

While the paragraph 'encouraged' ECOWAS to work towards a peaceful restoration of Kabbah's government, this, it could be argued, was only advice. Indeed, the first part of the paragraph, giving strong support to ECOWAS's whole three-pronged approach, was clear and absolute. Therefore, with the Security Council acting under Chapter VII of the UN Charter that only allows resort to force in certain circumstances,[14] it can be said the Resolution was not without ambiguity, whether intended or not.

At the same time as there was confusion over the use of force, there was equal confusion over which parties to the conflict the Security Council Resolution 1132 paragraph 6, covering the arms embargo, applied to. These ambiguities became the key controversies over Sandline International's role. The paragraph reads as follows,

[The Security Council] decides that all states shall prevent the sale or supply to Sierra Leone, by their nationals or from their territories, or using their flag vessels or aircraft, of petroleum products and arms and related matériel of all types, including weapons and ammunition, military vehicles and equipment, paramilitary equipment and spare parts for the aforementioned, whether or not originating in their territory.[15]

By referring to Sierra Leone, and not the parties involved in the conflict, the Resolution could be construed in a number of different ways, as was later the case. The British government argued, however, that it applied to all parties to the conflict and that the scope was geographical instead of being specific. As the Legg Report makes clear, 'there was no exception for sub-regional forces as in the ECOWAS embargo'.[16] On the other hand, the UN Assistant Secretary General (Legal Affairs) argued that, while ECOMOG did not benefit from an explicit general exemption under paragraph 6, it must enjoy an implied partial exemption for the purposes defined by the Council in the resolution since any other interpretation would lead to a paradoxical situation in which the Council, while entrusting ECOMOG with important responsibilities, at the same time deprived it of the means to carry out those responsibilities.[17]

This position was, of course, contrary to the FCO, in that the UN legal advisers understood the Resolution in the context of the overriding aim of restoring the government of President Kabbah.

The British government's view of the meaning of the Resolution is explained in paragraph 7. The pertinent part of the paragraph reads as follows,[18]

7. Decides that the Committee established by below may authorise, on a case-by-case basis under a no objection procedure:
    (a) applications by the democratically elected Government of Sierra Leone for the importation into Sierra Leone of petroleum products;

Since part of paragraph 7 is concerned with the oil embargo, referring to 'applications by the democratically elected Government of Sierra Leone', and, as such, distinguishing the Kabbah government, this could also imply that the reference in paragraph 6 to Sierra Leone is, indeed, geographic.[19]

Even so, the drafting is not at all clear. Vice Admiral West, a Ministry of Defence official, spoke of confusion about the meaning of the resolution, while President Kabbah believed that the Resolution did not apply to his government in exile.[20] Confusion over the meaning of the Resolution was further exacerbated by the way in which the FCO glossed over its meaning to its own Parliament, the media, and the exiled government of President Kabbah. The official telegrams announcing the Resolution and a number of FCO daily bulletins all referred specifically to the Junta and not President Kabbah's government.[21] Even the Edinburgh Communiqué issued at the Commonwealth Heads of Government Meeting (CHOGM) in October 1997 referred specifically to the embargo against the Junta. Furthermore, on 12 March 1998 The Minister of State, FCO, Mr Tony Lloyd, stated quite clearly during a Commons debate on the subject that '[the British government] were instrumental in drawing up the UN Security Council Resolution 1132, which imposed sanctions on the military Junta', while no mention was made of the Kabbah government.[22] A number of reasons have been put forward as to why doubts surrounded the meaning of the UN Resolution and that might have been responsible for it being played down by FCO officials. Sir Thomas Legg suggested in his report that the official policy of drying up arms to all sides 'was not published abroad ... because of sensitivities about the possible role of ECOWAS which, unlike HMG, had explicitly contemplated the use of force'.[23] The other reason was because '[British officials] genuinely saw the embargo as primarily aimed at the junta'.[24]

The final difficulty in implementing the UN arms embargo came through the Promulgation of the Order in Council, which caused further confusion among the relevant parties to the affair. As is detailed below, the result of this further confusion possibly strengthened Sandline's conviction that they were acting within the boundaries of the law. The Order in Council is implemented under Section 1 of the United Nations Act 1946. It was Mr Lloyd who, on 16 October 1997, approved the draft Order in Council, together with a second Order in Council covering the UK's dependent territories. They were formally made at a meeting of the Privy Council on 30 October 1997. The following day, they were laid before Parliament before coming into force on 1 November 1997.[25] They made it an offence, punishable by up to seven years' imprisonment to:

- supply or deliver;
- agree to supply or deliver;
- do any act calculated to promote the supply or delivery of;
- supply any of a long list of arms and military equipment to any person connected with Sierra Leone.

Sierra Leone was defined in the Order in Council as:

- the Government of Sierra Leone;
- any other person in, or resident in, Sierra Leone;
- any body incorporated or constituted under the law of Sierra Leone;
- any body, whether incorporated or constituted, which is controlled by any of the persons mentioned in sub-paragraphs (a) to (c) above; or
- any person acting on behalf of any of the persons mentioned in sub-paragraphs (a) to (d) above.[26]

As the Second Report points out, while the UN Security Council Resolution did not define Sierra Leone, the Order in Council did, thereby creating a significant pitfall for anyone inside or outside the FCO who read the UN Resolution but not the Order in Council.[27] With British law apparently clear, while the Security Council Resolution was ambiguous, it was also possible for a participant to argue that the Order in Council went beyond those measures called for by the Security Council Resolution.

ECOMOG actions

ECOMOG had been sanctioned to use force to intervene in Sierra Leone at the ECOWAS Heads of State Summit in Abuja. From the time of the Conakry Agreement signed in October 1997 until Kabbah's return to power in March 1998, ECOMOG began to build up a significant military presence at Lungi and Hastings on the east side of Freetown. ECOMOG's ground commander during this period was Colonel (later Brigadier) Maxwell Khobe who was in charge of ground operations throughout the offensive. At the same time as ECOMOG was gearing up for battle, the AFRC/RUF continued to bring in weapons and ammunition over the border from Liberia, while also delaying the start of the disarmament process agreed in the accord. After ECOMOG troops were attacked around Freetown and Lungi, the force commander went on the offensive against the AFRC. On 5 February, ECOMOG moved against the Junta. The battle raged for a week as ECOMOG forces advanced from Hastings and Jui on the eastern outskirts of the city, moving westwards towards the centre. After several days of heavy fighting, the Junta retreated from Freetown.[28]

The role of Sandline International

Sandline International was created in 1996 and registered in the Bahamas. The company was set up to supply military and security services to legitimate, internationally recognised governments, and relevant multinational organisations operating in high-risk areas of the world.[29] The company's Chief Executive at the time of the 'arms to Africa' debacle was Lt Colonel Tim Spicer, OBE, Retired, while Michael Grunberg, a chartered accountant and management consultant, acted as Sandline International's financial and business adviser.[30]

At the time of the affair, the company did not believe supplying weapons to Kabbah's government in exile was unlawful. While the shift towards legally acceptable customers did not start with Sandline International, the company reinforced the shift through its operating principles. Whereas a defining feature of classic mercenaries is their willingness to work for whoever will pay them the most, whatever the legal position of the client, Sandline International is very particular about the projects it accepts. The policy of the company is to ensure that they only accept projects, which, in the view of the management, would improve the state of security, stability and general conditions in client countries. To this end, the company would only undertake projects that are for:

- internationally recognised governments (preferably democratically elected)
- international institutions such as the UN
- genuine, internationally recognised and supported liberation movements.

and which are:

- legal and moral
- conducted to the standards of first world military forces
- where possible, broadly in accord with the policies of key western governments
- undertaken exclusively within the national boundaries of the client country.

Sandline would not become involved with:

- embargoed regimes
- terrorist organisations

- drug cartels and international organised crime
- illegal arms trading
- nuclear, biological, or chemical proliferation
- contravention of human rights
- any activity which breaches the basic Law of Armed Conflict.[31]

Furthermore, the way Sandline International openly conducted its business activities to do with Sierra Leone, strongly suggests that the company did believe they were acting according to the law and had the support of the government.

It was shortly after the Heads of State Summit in Abuja that Kabbah first made contact with Sandline International. According to Hirsch, the company 'sought to assist the Kabbah government in creating a military force to repel the AFRC/ RUF Junta from Freetown and the Kono diamond fields'.[32] But it was not until December that Kabbah had coordinated with the Nigerians in obtaining Sandline International's assistance for enhancing their combat ability. By January, the Civil Defence Force had carried out the first phase of the company's plan as they isolated AFRC/RUF forces in Bo and Kenema.[33] The next phase of the operation was late because of the delay in shipping out Bulgarian arms to Sierra Leone. On 22 February, Sandline International arranged, through Sky Air, to fly the arms out to Sierra Leone.

The weapons consignment consisted of about 1,000 AK 47s, mortars, light machine-guns, and ammunition. The company purchased the weapons from Bulgaria, and then had them flown to Lagos, where the aircraft was refuelled with ECOMOG's assistance for the next sector to Lungi airport. On arrival, the weapons were handed over to Nigerian peacekeepers and not distributed till later. At the same time, the Nigerians decided to replace some of their own older equipment from the shipment. According to a Nigerian military official, they were eventually handed out with about 5,000 weapons taken from both the rebels and loyalist militias.[34] Grunberg explained that the company had no detailed knowledge of what happened to the weapons after ECOMOG took control of them. He went on to stress that the weapons were not smuggled into Sierra Leone but imported with the knowledge of the government and the Nigerian authorities, i.e. ECOMOG. They were not seized by the Nigerians but taken into storage, the details of which are discussed in the analysis.[35]

The cost of Sandline International's operation was going to be met by Mr Rakesh Saxena, a Thai businessman then operating from Vancouver. Saxena indicated that he represented a group of investors with mining interests in Sierra Leone.[36] After further contact between President Kabbah and Saxena, both entered into two linked but separate contracts on 23 December 1997. The first contract was a memorandum of understanding between President Kabbah, on behalf of the Republic of Sierra Leone, and Mr Saxena, on behalf of the Blackstone Capital Corporation of Belize. Under the memorandum, President Kabbah granted certain mining exploration concessions in Sierra Leone in return for economic and other assistance to the value of US$10m to assist in the reconstruction of Sierra Leone.[37] The second contract was a supplementary memorandum of understanding, entitled

'Military Annex', and was between President Kabbah and Mr Spicer on behalf of Sandline International.[38] Saxena agreed to pay the funds he had committed to the government of Sierra Leone direct to Sandline International as a matter of expediency. Originally he had committed US$10m to the operation. He then dithered in making the payments according to Grunberg, who flew to Vancouver and secured US$1.5m from him and a promise of a further US$3m or US$3.5m to follow shortly and the rest later. In the event, he only paid the first instalment and Spicer had to modify his overall operational budget accordingly.[39]

The method Sandline International used to broker the weapons was not the typical method used for illegal transfers of weapons. At no stage did the company appear to hide the transfer of weapons from the UK government. For example, the company did not use a broker, which would have been the obvious way to hide such a transaction. There was also no indication that the company attempted to exploit loopholes and inadequacies in national and international arms control regulations. According to the Small Arms Survey 2001, brokers often exploit discrepancies between national arms control systems, taking advantage of inconsistent documentation requirements, and ineffective verification mechanisms. It is common, for example, to falsify details on end-user certificates to hide the true nature of the consignment of weapons and its final destination.[40] In the case of Sandline International, officials in the UK, US and Canada, where Saxena was based, were briefed on the company's intentions and the weapons component,[41] while details on the end-user certificate, which had been signed personally by President Kabbah, were in order, according to the Legg report.[42] Given the charge against the company that they acted illegally in exporting weapons to Sierra Leone, a close inspection of the end-user certificate would have been a priority of any investigation. While a detailed inspection may have happened, there is no indication from the Legg report that there were any irregularities about the end-user certificate. Instead, the way the purchase was conducted suggests the company believed they were acting within the law.

The way the weapons were transferred was also different to how illegal weapons transfers are normally conducted. Illegal arms transport flights, because they are illegal, will be secret. Everything possible is done to hide the true destination of the weapons. These flights rarely fly direct to their destination, but prefer circuitous routes involving multiple landings, refuelling, and changes of aircrafts.[43] Carrying out a typical delivery will involve several interacting groups of intermediaries and fellow collaborators spread over several countries. No evidence from the Legg report points to Sandline International acting this way. The flight route taken by the weapons, from Bulgaria to Sierra Leone *via* Nigeria (the principal member of the ECOMOG deployment), was never concealed, while Nigeria is not a country normally used to supply illegal weapons into Sierra Leone.[44]

The investigation

The investigation into whether Sandline International supplied arms to Sierra Leone in breach of the arms embargo was announced on 18 May 1998. The terms of the investigation were made in light of the allegations that the government was involved with the supply of arms to Sierra Leone. The investigation was to establish:[45]

- what was known by government officials (including military personnel) and Ministers about plans to supply arms to Sierra Leone after 8 October 1997;
- whether any official encouragement or approval was given to such plans or such supply; and,
- if so, on what authority.

According to the government the investigation was independent, even though it was run by a former senior civil servant and used government offices. Sir Thomas Legg also had access to all Government files, papers, and records relevant to the investigation. The investigation had the co-operation of HM Customs and Excise, as well as having access to officials and ministers. The period under investigation was from 8 October 1997 to 24 April 1998, when the Berwin Letter, setting out the names of the UK and US government officials that Sandline International asserted they had briefed in advance on their plans, arrived for the attention of the Foreign Secretary.[46] The letter, expressing concern over the Customs investigation against Sandline International, wanted to know from the Foreign Secretary why the company was being investigated for breaching an arms embargo which did not apply to Kabbah's government, and that British government officials were aware of.

The investigation received evidence from a wide range of sources. They included interviews with seven Ministers, two advisers, in some cases more than once, and 49 officials who also provided the investigation with summary statements of their involvement with the matter. Evidence was also received in the form of statements from three other Ministers and six other officials, who were not interviewed. The investigation also received evidence from Sandline International, in particular from Spicer and Grunberg. Through their solicitors, the investigation was allowed to see the full transcript of their interviews with Customs, together with other relevant documents.[47]

The investigation drew on official documents relevant to the matter. Great care was taken to ensure all relevant documents and records were produced so as not to undermine the scope of the investigation. A total of 123 key documents were listed as being of particular importance to the investigation. In addition to official documents, the investigation had access to 185 telegrams that passed between the FCO and the High Commission to Sierra Leone and between the High Commission and Abuja on the subject of Sierra Leone. Finally, all the intelligence reports and assessment relevant to the investigation were made available. The investigation saw 102 intelligence reports and assessments on Sierra Leone during the period May 1997 to May 1998. These reports are secret, and therefore cannot be made public at present. The investigation was satisfied that, with only one exception, they contained nothing that was significant to the investigation.[48]

Evaluation of the affair

The changing identity of PMCs lay at the root of the confusion over Sandline International's role in the Sierra Leone debacle as a result of the changing nature of war. The realisation by officials and MPs that PMCs are likely to play an increasing role on the international stage has led to them taking the necessary steps to examine more closely what these companies represent. It was not only the Sandline International affair that challenged the image held by the FCO that PMCs were nothing more than classic mercenaries with a corporate veneer. A number of other factors most certainly influenced their minds and thus accounted for the confusion over what Sandline International represented: classic mercenaries or something very different. The inability of officials to grasp the shifting nature of war, away from a state-centric activity conducted solely by state militaries to wars undertaken for economic or ethnic reasons, might have also contributed to FCO officials failing to recognise the shift in the role of PMCs in such conflicts. These wars involve a whole range of actors, including PMCs. The flexible nature of PMCs makes them well suited to these wars, while the impact of this shift has been the need for new policies and regulations to control PMC activity.

For the FCO to have recognised, and then adapt their thinking quickly to the changes in the way war was now being fought, especially in places such as Sierra Leone, was going to take time. Instead, FCO officials still understood war in largely conventional terms, fought between state militaries for political motives. The Sierra Leone war was being fought for economic and ethnic reasons. As such, the war represented a blurring between war – usually defined as violence between states or organised political groups for political motives purpose; organised crime – violence undertaken by privately organised groups for private financial gain; and large-scale violations of human rights – violence undertaken by states or politically organised groups against individuals.[49]

The company, however, had different ideas about their status. They saw themselves as legitimate actors, who were supporting a democratically elected government back to power. Furthermore, the company was in a position to achieve this with minimum cost to the British government. In effect, while Spicer and Grunberg had identified an opening in the market, which they were intending to fill as legitimate actors, the FCO had different ideas, but perhaps in large part because they had not yet responded to the changing nature of war in the same way Sandline International had done.

The ambiguity of the FCO's position as a consequence of the historical role of mercenaries in UK foreign policy position also helped support the idea that the company was different to that of the classic mercenary, and may also explain the reason why the FCO was communicating with them in the first place. Admittedly, some officials argued that they did have reservations about this, though it is impossible to say whether such reservations were made in hindsight to cover themselves, or that they truly existed prior to the events in question. Notably Everard and Murray, both of whom were responsible for looking after the FCO's policy over West Africa, and, as such, were caught up in the 'arms to Africa' affair.[50] It may be that individuals were confused over what PMCs represented, a mercenary company or a legal business, though this is unlikely. It is more probable that there occurred a clash between traditional secrecy associated with mercenaries with the new legality now associated with PMCs, while reinforcing this was the different way Sandline International was treated by the FCO from how mercenaries were treated by the government during the 1960s and 1970s. As discussed in Chapter 2, the government had always kept mercenaries at arm's distance, so as to be able to officially deny having anything to do with them.

On the other hand, FCO officials, who stopped short of accepting Sandline International as a fully legitimate organisation able to contribute to its efforts, were able to engage in open but guarded discussions with the company's management. As both the Legg Report and the report by the Foreign Affairs Committee noted, both parties entered into dialogue, even if the contents of that dialogue have been contested.[51] The fact was, Sandline International openly conducted discussions with the FCO, even though, as Mr Everard, deputy head of the FCO's West African department, explained to the Legg Committee that, 'he was instinctively wary because of the possible risk of allegations of dealing with mercenaries'.[52] Such allegations would have invariably come from the press keen to exploit a good story, or organisations with a political agenda that challenged the right of companies to privatise certain wars, as Sandline International appeared to want. Even so, it can be argued that by conducting a transparent dialogue with Sandline International, which was essential to the company to ensure they were not crossing government policy or intent, the company's legitimacy benefited even if this was not the company's intention.

The argument can even be extended. Through the talks they held with Sandline International, the FCO would have known about the company's operation, though not necessarily the part where the company was to supply arms to the Kabbah government, though Sandline International disputes this, maintaining they did tell the FCO and the US State Department. After all explains Grunberg, why would we be selective in what we told them as this would only reflect badly in the future.[53] The failure of the FCO to stop the operation, even without knowledge of weapons, as soon as they became aware of it suggested to the company that they were acting legally in the eyes of the government.[54] Similarly, not intervening also suggests they gave a level of approval to the company's actions in support of the Kabbah Government. As such, the company's management could reasonably assume their actions were legitimate and endorsed.

As a result of the confusion in the FCO over what the company represented, Sandline International entered into two contracts with Kabbah as mentioned earlier.[55] This decision was itself based, in part, on the belief that Sandline International was doing nothing wrong and had the support of the FCO. The company did not see anything immoral in conducting an operation with what they thought was the full support of the British government to restore to power the democratically elected leader of the country. As is shown below, there was sufficient ambiguity in the dialogue for both sides to believe they were acting in an appropriate manner, and, as such, were doing nothing wrong. In the case of

Sandline International, they believed their actions to be legal because they thought they had approval and support from the High Commissioner[56] and senior civil servants on the Africa desk, and therefore approval from the FCO, and ultimately the British government, when in fact no Ministerial approval existed. At the same time, the insufficient knowledge of FCO officials on the subject of PMCs also suggests there was a level of confusion as to what the company represented.

Bringing the empirical and thematic together

What themes, using the empirical evidence presented throughout the chapter, did the affair highlight, and, as such, need to be analysed in relation to the significance on the legal status of PMCs today? The first theme has to do with UN Resolution No. 1132. Was there any confusion among the different parties helping resolve the conflict as to who the resolution applied to? The answer is yes. Following on from here, the next question we need to ask is whether the company was entitled to supply weapons in the first place. They certainly thought so. Again, the open manner in which the arms transfer occurred suggests the company thought its actions legal and above board. Finally, what about the confusion experienced by FCO officials over Sandline International's identity? The way FCO officials handled their communications with the company was very different to the way they had dealt with mercenaries in the past; an indication that some type of legal shift was occurring, but had not been fully appreciated by officials in the FCO.

ECOWAS had made their intention to use force quite clear in a policy statement, and it could be read to include the purchase of weapons. Elsewhere, there remained what appears to be confusion over the nature of the role of Sandline International as either arms dealers, military advisers, or both. This is not surprising given that military services are often accompanied by the sale of weapons,[57] as Sandline's contract stipulated in Sierra Leone. Anderson[58] also recognised the problem of distinguishing between both sets of activities when he asked questions about the UN Security Council Resolution 1132 during a debate on Sierra Leone,

> It may be that Sandline was initially supplying a different sort of assistance to the ECOMOG forces. It might have provided technical assistance, intelligence and training, in which case it would not have been in breach of the resolutions [such activity could be termed as mercenary in nature], which is essentially a ban on arms supply. It may have been like the United States forces in Vietnam, who began as military advisers, but gradually slipped into a combat role.[59]

It is worth noting at this point that the weapons supply was only one component of the overall serviced provision in support of ECOMOG forces. The company supplied a range of other services and activities encapsulated within the contracted strategic objectives. They were involved in the ECOMOG planning group at the highest level, while the company also provided communication equipment for President Kabbah's government in exile. The Sandline helicopter was used to deploy numerous UN, NGO, and US government reconnaissance parties trying to provide aid and support to those areas inaccessible by road. In addition to this, the helicopter ferried ECOMOG troops and equipment to their destinations, evacuated wounded soldiers and civilians, distributed humanitarian aid packages to remote regions, and rescued numerous local and expatriate workers who found themselves in rebel held areas. If they had not been rescued they would most certainly have been killed. The company also provided tactical intelligence to UN and US personnel and detailed advice and tactical briefings to Royal Navy helicopter pilots. Finally, the company protected strategic and commercial assets from looting and vandalism throughout the war.[60] None of these tasks can be said to be insignificant and yet the focus of the investigation was solely the supply of weapons.

Did Sandline International operate within the boundaries of the law? The transparent way the weapons were transferred, not hidden from official scrutiny by British or other international officials responsible for monitoring the arms embargo, would indicate the company believed their actions in importing weapons into Sierra Leone on behalf of the Kabbah government were legal and undertaken with the prior knowledge of UK government officials. Why would the company act the way it did act, if it thought its actions were illegal? Instead, the way the company undertook the weapons transfer suggests it did not know it was purportedly technically breaking the arms embargo, but instead believed the British government was endorsing its actions. The British government, on the other hand, saw the shipment of weapons as a potential breach of the arms embargo. According to them the company acted outside of the letter of the law in supplying weapons to Kabbah's government. Consequently, responsibility for the debacle should not lie solely with the company. The FCO was also to blame for the debacle by not making themselves familiar with the Order in Council that prohibited the export of weapons to Kabbah's government and making this known to Sandline officials when they sought advice, while no one had highlighted the nature of a law that prevented the external provision of essential assistance to a democratically elected party that had been ousted in a military coup.

With respect to the High Commissioner, the lack of effective knowledge about the Order in Council meant he gave Sandline International's plan a degree of approval, and in doing so may have overstepped his authority. He contends, however, that he kept the FCO advised in a written notice sent at the beginning of February of Sandline International's plan, but never received any instructions telling him to act otherwise.[61] Neither did he know such a shipment was illegal,[62] while other FCO officials were also ignorant of the fact the arms embargo prohibited the supply of weapons to all parties. In dealing with Sandline International, FCO officials were conscious of the need for a degree of care in dealing with PMCs; though it is quite possible it was the event itself which made them nervous – a consequence of the pejorative pre-conceptions attached to the business activities of the company. Everard recognised that in the ground rules on which he conducted his conversations with Mr Spicer.[63] These rules, which according to Grunberg were all stated after the event, included not taking the initiative in making contact, limiting contact to telephone conversations, avoiding any comment which might suggest approval or disapproval of (the company's) activities, and noting information provided by Mr Spicer, but not directly soliciting it,[64] while no one in the Africa department thought about taking full minutes of the meetings till after the event.[65] None of this, however, stopped FCO officials from having contact with the company either directly, or by phone.[66] The nature of the contact between the FCO and officials from Sandline International suggests, at the very least, that FCO officials were unsure about the true nature of this new organisation, therefore suggesting some type of shift in legitimacy was occurring at this time, but was not recognised by FCO officials.

Was the company entitled to supply arms to support ECOMOG actions? According to the Order in Council, they were not entitled to supply weapons, even though the UN legal office argued that the arms embargo imposed on the country contained an 'implied exemption' for the arms shipment to ECOMOG.[67] Unfortunately, the company failed to realise this, as did government officials. Was Sandline International responsible for this failure? Since the company had no idea about the existence of the Order in Council they can hardly be held responsible for failing to realise that supplying arms to support ECOMOG was a breach, except that ignorance is no excuse in the eyes of the law. Should Sandline International have known about the Order in Council? Given that a number of FCO officials directly involved with Sierra Leone did not know of its existence or appreciate its content, one can hardly expect officials from Sandline International to have known of it. And even Ministers misinterpreted it in public statements by narrowing its application to 'the Junta'. Finally, one would be entitled to expect the High Commissioner to advise Spicer of the Order in Council at the briefing meetings they had if he had known of its existence.

The fact that the arms transfer was transparent suggests the company believed they were acting in a law abiding way. If the company honestly believed its actions were illegal and were prepared to work outside the law, and there is no evidence to suggest this was so, they would have acted differently in order to camouflage the supply of arms, or sought to obtain an explicit licence under the DTI regulatory framework.[68] Neither would the company have openly flown the weapons directly from Bulgaria *via* Nigeria, the country in charge of ECOMOG, and thus receiving assistance from Sandline International, if they knew their actions were illegal. Instead, the way the company acted suggests they were ignorant of, or operating with a mistaken interpretation of, any law relating to the UN Resolution being broken; and in any event believed they had the full endorsement of the British government. Nor were the details into the arms transfer ever investigated by Legg. Considering Legg was investigating what Ministers knew of the illegal transfer of weapons to the Kabbah government, and whether any British official gave approval when they were not sanctioned to do so, it is strange no mention is made of the transfer in the investigation's report. It may have been left out because it confirmed what the company had said all along, that no one really knew they were potentially breaking the letter of the law, while adding to the list of questions already facing the FCO.

Did the Order in Council go beyond the measures called for by the Security Council Resolution? This argument may well have been heard if the Sandline case had come to court,[69] though Sir Franklin Berman, the chief FCO legal adviser, contested this position when he wrote that 'in implementing arms embargoes in domestic law, it is standard practice to include provisions which will enable the embargo to be effectively applied',[70] and thereby suggesting a degree of latitude, albeit in this case the embargo covered a wider target group than the original Security Council Resolution.

Finally, the FCO lacked any proper administrative procedures for promulgating an Order in Council of this kind, including explanatory material that may be necessary. Nor was the Order in Council given full publicity, the Order being seen as a technical formality of little significance to the general public.[71] Spicer also commented on the lack of publicity,[72] while the Second Report noted how, 'it [had been] treated in a disgracefully casual manner'.[73] Even more curious was the finding by Sir Thomas Legg that 'most of the main players ... have affirmed that they had either no, or only a vague and very general, awareness of the existence of the Order ... '.[74] For example neither Ann Grant, who was the FCO's Head of Africa Department, Equatorial, or Peter Penfold, Britain's High Commissioner to Sierra Leone at the time, saw the Order; in fact, the FCO stated that 'until recently, it was the practice to circulate copies of the relevant resolution, rather than copies of the Order in Council, or summaries of them, to the relevant High Commissions or embassies'.[75] Neither does it appear that the territorial departments in the FCO concerned were consulted fully during the drafting of the resolution. One of the recommendations of the Foreign Affairs Committee was that the,

> territorial departments concerned within the Foreign Office should be consulted fully during the drafting of any future British legislation which affects any specific foreign territory, and that, when the legislation is finalised, its text, and any necessary explanatory material, is brought to the personal attention of all Foreign Office officials who deal with that territory, both at home and overseas.[76]

Nor was the Order in Council subject to any parliamentary procedure. If it had been, the Order might have had to be defended on the floor of the House of Lords, giving wider attention to its meaning.

The very poor handling of the Order of Council was the last mistake in a catalogue of errors that ended in a debacle for the British Government. Confusion, ambiguity, and poor publicity were all to blame for the true meaning of the Order not being made apparent to all necessary parties, including Sandline International. Whether, as was suggested by Sir John Kerr, Spicer should have known the law in respect of the Order is questionable, more so since FCO officials themselves were not clear on the exact legal position of the embargo.[77] But, then again, the company's lawyers should have been aware of the Order, and been able to explain it. But, if Spicer really did believe Sandline International was acting within the law, why involve them? Spicer is not the only person who has carried out an action believing it to be within the law, when in fact it was not. Finally, as with the problems over the use of force and the UN resolution, the problems to do with the Order were another set of pieces in the jigsaw that has led to questions about the legal status of PMCs. This then, was the position on 30 March 1998 when Customs and Excise officials, believing they had *prima facie* evidence of a breach of the embargo, instigated an investigation, raiding Sandline's office and, although not widely reported in the subsequent press coverage, the FCO at the same time. They also met with Michael Grunberg of Sandline International that same day.[78]

Moreover, at the institutional level, and not at the level of individual officials, the FCO's denial of knowing anything about Sandline International supplying weapons to ECOMOG was normal in light of the institution's historical relationship with mercenaries, who, as illegal combatants, are not entitled to supply weapons or any type of military service to sides in a civil war. The collective memory of the FCO simply failed to recognise that Sandline International was not a mercenary outfit, but a legally registered company, which had every right to enter into a contract to supply weapons to ECOMOG. Only now is the distinction between these two groups gaining ground. For the FCO not to react in the way it did, but to keep quiet, could have delayed the process of legitimating PMCs even further, by delaying the Green Paper, which emphasises their utility. In this respect, the government found itself obligated to recognise the growth of the sector and consider a regulatory framework purely as a result of the focus generated by the 'affair'.

At the same time, the parliamentary debates on the procurement of arms for Sierra Leone[79] further helped the company's image, since other aspects of the company's business, essentially those activities closely associated with mercenarism were marginalised. In this respect, the tone of these debates, while criticising the company for its actions, did not extend to a full rejection of PMCs, as would have happened if Sandline International had been perceived as a mercenary outfit from the outset of the investigations; although the company was effectively ostracised by the government almost as if a grudge was borne against it and which was severely detrimental to the firm's image according to Grunberg.[80]

Sandline International was certainly not seen as classic mercenary outfit, since it would be fair to suggest that if they had been, the FCO would have terminated any open contact with them immediately. By not doing so, a level of confusion by FCO officials over what Sandline International represented was implied. This should not be seen just as a failure on the part of the FCO. Instead, it is the result of the industry's rapid development especially in terms of its legal status, and which individuals in the FCO may not have been aware of, or understood, at the time.

A further reason why FCO officials may have been confused over the identity of Sandline International has to do with the institutional nature of the FCO and which exacerbated the whole affair. As suggested above, the company did not fit the normal profile of a mercenary outfit that had operated during the 1960s and 1970s. At the same time, for the FCO to identify the company as something very different to classic mercenaries was not going to be easy institutionally at an early stage of the development of PMCs.

All these problems were later examined in the government's Green Paper on 'Private Military Companies: Options for Regulation'. As for the tone of the Green Paper's response to PMCs, it was positive. According to Grunberg, they recognised such companies could make a contribution to British foreign policy.[81] This, in turn, saw the government's original position on PMCs shifted. They now accept there may be a limited role for PMCs in peacekeeping operations.

The transformation of the legal status of PMC activity

It is important to understand exactly what is being transformed here. What is not being transformed is the legal status of PMCs as companies: PMCs, such as Sandline International, are legal business organisations, registered at Companies House or an equivalent authority overseas. Instead, we are interested in the activities PMCs undertake, in particular how the legal status of these activities is changing. The transformation of the legal status of any activity could be said to occur when the legal position of the activity changes, either moving outside the law or moving within the law. For PMCs, this transformation has to do more with society's perception of the type of activity acceptable for PMCs to undertake. That such a transformation is occurring is not in doubt. How else can we account for the changes in attitude that have occurred over the last few years? Today the public accept the increasing number of military orientated roles undertaken by PMCs, including training foreign military forces, or supplying logistical support to military operations. Yet, such activities undertaken during the Cold War would have been associated with mercenaries, bringing into question the legality of PMCs. How has this transformation occurred?

The best evidence of the changing perception of PMCs and the growing realisation of PMCs becoming a permanent feature of international security was the publication of the Green Paper in 2002, a delayed consequence of the Sierra Leone debacle in 1998. Consequently, the need to resolve the problems of identity has taken on a greater urgency. By publishing this paper, which broadly accepts the need for considering a licensing regime able to distinguish between reputable and disreputable private sector operators,[82] encouraging and supporting the former, while as far as possible eliminating the latter, the government was in effect accepting and acknowledging changes to the way society understood certain types of non-state violence and its role in the international community.

Contrast this position towards a PMC with the position of the government in 1976 towards British mercenaries who had fought in Angola, and were then put on trial after their capture. Then, the only action taken by the government was to order a report into their activities, before quietly erasing the whole issue from the public arena. By not taking the issue any further, the government recognised that the general public was not ready, or prepared, to compromise their principles to do with the nature of mercenary activity. Restricted because of the principles held by the general public, it became impossible for the government to think about changing the legal status of the activities of mercenaries, as is happening now.

The significance of the 'arms to Africa' affair

Though not necessarily realised at the time, the 'arms to Africa' affair was probably the turning point for commercial security companies determined to legitimate their role in the area of combat and combat support operations. As a result of three investigations, one of which was by the parliamentary Intelligence sub-committee, and the subsequent debates on the subject in the House of Commons, conventional understanding on the role of mercenaries in international security is now being revised. Before, when mercenaries had undertaken combat and combat support operations, it had normally been understood that these roles were illegal,[83] or if not illegal, certainly unpalatable even when occasionally consented to by a government with a vested interest in the operation.

Sandline International is not the only PMC to recognise the importance of the shift in the legal nature of PMC activities over the last decade. Without the shift, it is doubtful whether the industry would have made the type of progress it has over the last few years. The shift was essential if the industry was to move forward. On the other hand, it has not been as smooth a shift as some in the industry would have liked. The 'arms to Africa' affair has highlighted the type of problems that can arise from the shift unless regulation is brought in to control the activities of the industry or the industry is able to engender wider confidence by establishing a transparent and credible self-regulatory environment. Even so, a shift in the legal status of PMC activities has occurred, and which the 'arms to Africa' affair has helped emphasise.

Furthermore, while the affair challenged conventional ideas on the role of PMCs in war, it is intended for the Green Paper[84] to establish policy options that will enable the government to direct PMCs into an area of work that can best suit their skills.[85] This process is still ongoing. In this respect, the present debate is unique, in that no other government has yet openly debated the role of PMCs in international security before deciding their function. Instead, policy options have either been determined through national legislation, as in the case of South Africa and the US,[86] or simply ignored.

Conclusion

The British public has generally displayed a dislike towards mercenaries, even though they appear to love the romanticised version portrayed in popular movies such as Wild Geese. This loathing now appears to include those who work for PMCs, who they see as representing a new manifestation of mercenarism. This is in part due to a conditioning by press reporting, and a lack of distinction therein, while the recent Equatorial Guinea affair has not helped matters. At the time of the 'arms to Africa' affair, this objection was reflected in the statements made by government officials against the company. Under these conditions, introducing legislation to outlaw, or to tightly control, the activities of PMCs would have been a simple matter of procedure. But, as pointed out at the beginning of the chapter, during the time it took to investigate the affair, and the publication of a Green Paper in February 2002, there occurred a significant, though subtle, shift in the perceived legality of the activities of PMCs and thus how the government came to understand PMCs as legal entities.

As discussed above, the shift in legality revolved around the activities undertaken by the company, in this case an arms transfer, not the supply of military services closely associated with mercenarism. Had the investigation by Sir Thomas Legg also focused on the military services the company were supplying, instead of the arms transfer, a very different outcome may have occurred.[87] It did not, and as a consequence, it was felt that while the company was in breach of the arms embargo it was not in the public interest to prosecute. Presumably, this was because it would expose the fact that the company was entitled to believe it had the tacit approval of HMG, as concluded by the independent Legg Inquiry, and endorsed by the Foreign Affairs Committee, according to Grunberg.[88]

Why was it that Sir Thomas Legg was not instructed to investigate the supply of military services, which included the provision of technical know-how, military logistics, and equipment,[89] by Sandline International to Kabbah's government? After all, the supply of military services is more closely associated with mercenarism, or was in the past, and may have revealed the company to be nothing more than a corporate front for the employment of mercenaries. The problem was that since the end of the Cold War military services have been supplied more and more by defence contractors exploiting the insecurity felt by many developing states, thus creating a market opportunity. At the same time, the provision of military services, as opposed to equipment, has not been something contemplated in UK law in the past. Trying to secure a successful prosecution under the present law by arguing that the supply of military services to ECOMOG represented nothing more than a mercenary activity would have been very difficult, particularly in light of the supply of such services by defence contractors to other states. Furthermore, Legg would have had to report that the service provision by Sandline International swung the war in ECOMOG's favour resulting in their victory, something the government may not have been keen to do, given their poor treatment of the company after the events became public knowledge.

The significance of the 'arms to Africa' affair is that it challenged the government's, and the general public's, understanding of the meaning of mercenary; a term that is vague and value-laden at the best of times. This challenge is still ongoing, and will continue for the time being. If the problem of definition is not resolved, the future for British PMCs in particular will remain uncertain, and, as a consequence, they will continue to operate in the shadows of the international community and kept at a discreet distance by the mandarins in Whitehall, while American PMCs strengthen their dominance of the market place and enjoy close working relationships with their government. Such a negative attitude is not categorical, but a symptom of our moral and political mindset.

And yet, this moral mindset is being challenged by a number of academics, multinational organisations, and government departments, as the Green Paper illustrates. The Green Paper highlighted the potential of PMCs as an alternative military tool, with economic benefits for the government.[90] This position may come to resemble the position African leaders found themselves in during the 1960s and 1970s, when they too came to rely on mercenaries, resisting any attempt to remove them from the international stage, while controlling their activities using international law.

Driving these changes has been the need for PMCs to legitimise themselves to a wider audience so as to be able to expand their business. Without some degree of wider support from the general population, which would be necessary for a government prepared to take the political risk of employing a PMC, business opportunities referred by or assisted by the UK government will remain limited to passive support, compared with the US where the government not only helps its PMCs acquire new clients through government-to-government relationships but is also the principal funder of their operations and considers them to be an effective tool in many international locations where it has an interest. More importantly, the lucrative sector of the market, in particular combat and combat support operations, will remain out of bounds. Finally, as a result of the 'arms to Africa' affair, the British government is now starting to recognise the need to take action to decide the future for PMCs.

The changing image of the mercenary through the ages is no accident. These changes reflect the privatisation of violence, but also changes in the international order such as the end of the Cold War. The same is true of PMCs. Indeed PMCs may represent a radically different image in 10 years' time. But, as with any process, change can be very slow. As the process of change occurs, leaving behind old images of the PMC, to take on board new images will invariably lead to a certain amount of uncertainty as to what the new represents, and whether it truly is a change from the past. Jack Straw, the Secretary of State for the FCO, also recognised this when he wrote in the foreword to the Green Paper:

> Today's world is a far cry from the 1960s when private military activities usually meant mercenaries of the rather unsavoury kind involved in post-colonial or neo-colonial conflicts. Such people still exist; and some of them maybe present at the lower end of the spectrum of the private military companies. One of the reasons for considering the option of a licensing regime is that it may be desirable to distinguish between reputable and disreputable private sector operators, to encourage and support the former while, as far as possible, eliminate the latter.[91]

What this statement makes clear is the need to re-examine the meaning of PMCs, and then explain where different types of PMC fit into the global security agenda. This is a significant step when any type of re-examination would have been impossible only a few decades ago, and is the focus of the next two chapters of the book. Even so, a re-examination and a wider acceptance at home must be attempted if we are to better understand the purpose behind the emergence of these actors and take advantage of the capabilities that they can offer.

5

THE PRIVATISATION
OF WARFARE AND THE
EMERGENCE OF PMCs

Introduction

The post-Cold War era is experiencing a revolution in how Western countries project military power. While war was once the prerogative of states, this is no longer the case. More importantly, our image of warfare is obsolete. Even so, state militaries are not being marginalised. Instead, their role in warfare is being transformed. The reason for such change is more to do with efficiency, technology, and shifting social perceptions of the role of state militaries in the twenty-first century. In the first Gulf War the ratio of US troops on the ground to private contractors was 50:1. By the second Gulf War that figure had increased dramatically to 10:1, as it was for the interventions in Bosnia and Kosovo.[1] A sense of the size of the shift can be gauged by the size of contracts awarded to companies working in Iraq. Halliburton's Kellogg Brown & Root (KBR) has been awarded contracts worth an estimated US$10.8 billion, while under the US Army's Logistical Augmentation Program contract, worth US$5.6 billion, the company can be called on by the Army to support all of its field operations.[2] Private contracts are also supplying more mundane services to the military away from the battlefield. These changes, however, have not developed at the same pace between state militaries. While the US military is starting to embed private contractors within its organisation, other state militaries in the West still have some way to go to catch up.

Neither has this change to warfare been confined to state militaries or simply government agencies, while outsourcing logistical support is only one of many services now undertaken by private contractors. As the following pages explain, governments are outsourcing responsibilities that in the past were associated with Foreign Service agencies, intelligence services, and development agencies. Even companies undertaking post-conflict reconstruction work are contracting companies to supply military/security services to protect their staff. Such is the growth in the industry that publicly quoted multinational companies are among the biggest suppliers of such services. However, it is not yet a market dominated by large corporations. Smaller companies, particularly in Europe, have also demonstrated their worth. In Iraq some of the largest security contracts went to small PSCs based in the UK. What is clear, however, is that the market for these services is truly global, raising some very fundamental concerns about accountability, transparency, oversight, and future trends.

This chapter seeks to explain the rationale behind the growth today in PMCs selling military and security services. In doing so, it first examines the reasons why the use of PMCs by governments is growing. Part of the explanation has to do with the reduction in manpower of state militaries with the end of the Cold War while other explanations focus on cost, the increasing complexity of technology on the battlefield and society's changing attitude to the application of lethal force in support of national interests. Following this discussion, the chapter focuses on those government responsibilities, logistics, intelligence, and force protection now being outsourced, and rationalises the need to contract out such services to the private sector. Next, the chapter attempts to show the extent of outsourcing and how it became a global phenomenon. It is not just militaries that use PMCs but international organisations, multinational corporations and even some international non-governmental organisations (INGOs). Nor are PMCs restricted to operating in a particular geographical zone. Instead they are to be found working on every continent. Finally, the chapter looks at future trends in the market for PMCs.

The forces driving the privatisation of war

The expansion of the PMC market is a result of a number of issues that have changed the way we now think about warfare. They are in part the result of a set of historical links with the past that has made outsourcing easier than if the link had not existed. As explained in an earlier chapter, UK PMCs are rooted in the Cold War, while in the US they can be traced back to the two World Wars and the military industrial complex (MIC). In the UK, PMCs are the successors of the private violence that was occasionally used to support the government's foreign policy during the Cold War, while the US government has a history of using the corporate world to assist its military perform their duty. However, other more recent factors are now responsible for growth of the industry.

Since the end of the Cold War, Western militaries have seen their numbers shrink. The US military alone has seen its numbers decline from 780,000 troops in 1991 to 380,000 today, fuelling the growth in private contractors.[3] The US military is not the only one facing declining numbers. Wide-ranging structural changes to the British Army will see the number of infantry battalions fall from 40 to 36 as the army struggles to recruit more people into its ranks. The intended cuts could see Britain with its smallest army since the Afghan War of 1839.[4] At the same time, the MOD is set to axe 20,000 posts in the next few years as part of a modernisation plan, while PMCs and PSCs will invariably benefit from such job losses. Other Western militaries face the same worries over recruitment, while they will also need to restructure to make themselves a more robust and resilient force that will be able to tackle the challenges of the twenty-first century.

The increasing reliance on technology by Western militaries, but particularly the US military, has made it very difficult for some sectors in the military to function without civilian support. Today, the military uses off-the-shelf technology adapted to military requirements. Maintaining this sophisticated technology requires a level of expertise beyond that which is taught in the military. Weapon systems, such as the Aegis missile defence system and Patriot missile batteries, along with unmanned aerial vehicles and Apache helicopters, to mention only some of the high technology equipment available to the US military, all need civilian technicians to run them. At the same time, other industrial countries are also likely to incorporate a US-style technological approach to defence, though not to the same extent as the US government. Neither is it unlikely the situation will be reversed any time soon, since the majority of research and development in fields such as Information Technology (IT) is undertaken in the market place not by the military, thus increasing the military's need for civilian technicians.

Such growth in military services is also the result of the shifting structure of modern society. During the Cold War, geographical conflict was the rationale for the existence of state militaries. Politics is still an essential component, but it no longer dominates our thinking as it used to do. Other issues more akin to present socio-political and economic pressures have raised questions within modern society as to whether state militaries are the only option available to governments conducting security operations. To a large extent, the increasing reliance on PMCs and PSCs reflects concerns about monitory efficiency and the application of lethal force to save strangers. In relation to the former, the introduction of neo-liberal economic ideas during the Reagan and Thatcher era was based on an assumption that the free market could deliver public services more cost efficiently. In light of this, many public services were sold off to the private sector. The military were not left out of the process and many non-core functions were sold to private enterprises to save taxpayers' money.

In the latter case, the concern over the use of lethal force to save strangers reflects an increasing determination by some Western governments to shape the global security environment, while accepting the reality that today the public will not tolerate troop casualties as it has done in the past. The body bag syndrome, as it is sometimes called, has increased the pressure on governments to find alternative means of carrying out some types of security operations, especially where national interests are not directly at stake. Finally, particularly in the case of the US, contracting out security operations allows governments to take the credit when things go well, but to avoid taking the blame should things go wrong.

Creating space in the military for PMCs

The mid- to late 1990s have proved a difficult period for state militaries trying to transform themselves to be able to respond to civil wars, low intensity conflicts, and humanitarian emergencies, while also preparing for conventional war. Prior to 11 September 2001, support in the West for humanitarian intervention was waning from the initial support it had received immediately after the Cold War. The experience of the US military in Somalia was a significant factor in souring attitudes towards foreign deployment to protect individuals without the means to protect themselves. The most obvious expression of this attitude was the Presidential Decision Directive (PDD) 25, which stipulated a long list of conditions for US intervention.[5] According to Howe, 'fighting and suffering in Somalia was still occurring in late 1999, but the West no longer expressed any willingness to assist in ending the conflict'.[6] Ignatieff is more direct when he argues that 'we are losing our capacity to do good in the world because we are no longer willing to risk the moral danger of doing evil'.[7] The Rwandan genocide revealed the swing against military intervention. The French and several allied African countries eventually sent troops to establish a safe area for fleeing refugees, though even this was criticised as an attempt by France to stop a Tutsi-dominated Rwandan Patriotic Front (RPF) from winning a clear and decisive victory in Rwanda.

The reduction in public support for military intervention has not led to the demise of the state military to be replaced by PMCs. That was never going to happen. While the social environment during the mid- to late 1990s was very different from that of the Cold War, the need for a state military remains; Afghanistan and Iraq are examples of that need. What we are seeing, however, is the creation of more opportunities for PMCs to operate within the military sphere as a result of the reluctance of politicians to commit soldiers to conflicts where there are no national interests at stake. And even when states do commit troops, many of the non-core functions on these deployments are carried out by civilian contractors. Thus, a division of labour is now occurring in the military itself. As one MPRI official put it, 'in the new downsized army soldiers ... don't do KP[8] anymore. We don't need to spend all that money and effort training a fine combat soldier and have him peeling potatoes'.[9] While soldiers will continue to fight, PMCs are taking on more and more non-core roles that do not involve actual combat, even though such roles can be crucial to the success of the combat mission. Without the support of private contractors some state militaries would now find it virtually impossible to fight. Contractors allow the military to concentrate on its core function, fighting wars, by removing responsibility for the more mundane operations, which are no less important to maintaining operational efficiency, and handing that responsibility to outside agents.

Two approaches to outsourcing military activities to PMCs

The roles taken on by PMCs tend to mirror the primary functions that their state military excels in. While PMCs can recruit from a global market, they remain tightly associated with their national contexts. At the same time, the ways PMCs have developed in the US and UK reflect the needs of their primary clients. As a result, the organisational structure of US and UK PMCs is very different. In the case of the US, multinational corporations, such as DynCorp and MPRI, undertake much of the work now associated with PMCs, while the majority of their work comes from US and foreign government contracts. UK PMCs are much smaller than their US counterparts, and draw the majority of their work from the commercial sector. As already mentioned, size is not necessarily an indication of competence and efficiency in the market place. As Iraq has demonstrated, UK PMCs may be smaller but this has not prevented them from biding for and winning some of the largest security contracts in the country and against very fierce competition from

US PMCs. The following two sections examine the different approach taken in the US and UK to outsourcing military activities.

The US approach to outsourcing technical and non-core military services

The history of outsourcing non-core military roles to US PMCs and US PSCs can be traced back to the early part of the last century. Their roots are located in the engineering sector of the US economy, where even today most of these companies are still located. This means their presence at the national level has been felt for some time, though not necessarily in the security sector. The industry has always had a close working relationship with the government, especially the defence sector, either supplying them with technical solutions or managing military assignments. This relationship grew closer with the introduction of computer technology during the Cold War. In this political environment, science was anything but academic, with a large part of the country's industrial energy going towards supporting the military. At the same time, defence contractors started to dominate military research and development.[10] The name eventually given to this relationship was the Military Industrial Complex (MIC). The MIC created, according to Leslie, 'a new kind of post-war science, one that blurred traditional distinctions between theory and practice, science and engineering, civilian and military, and classified and unclassified ...'.[11]

The MIC represented a working partnership between the military and civilian contractors. During the 1950s and 1960s, civilian technicians were a minor component of the defence establishment. Their numbers increased as the military came to rely on technically complex weapons systems.[12] But the range of military activities being outsourced, the actual intensity of the outsourcing, and level of integration between the military and companies has changed significantly in recent years. As discussed earlier, the application of market mechanisms to cut costs, reduction in the size of state militaries, increasing demands on the services of the military, and the availability of off-the-shelf civilian technology to replace military technology have all played a part in convincing government officials that privatisation of some military functions is inevitable. In future, the military may find itself focusing much more on improving operational capabilities by moving soldiers into military roles directly related to combat operations, while private contractors, including PMCs and PSCs undertake many of the rear echelon tasks. How easy, or how long the task will take, however, is not clear.

The nature of the relationship between the corporate world and the military could be described as of mutual benefit and spanning most of the twentieth century. As a consequence of their long involvement with the military, US corporations today supplying military support services are typically very large commercial enterprises. Such corporations employ hundreds, if not thousands of people, with million dollar turnovers, and operate out of offices that are close to the heart of government. The close proximity of their headquarters to the institutions of government is reflected in the close working relationship between the two groups. These corporations supply the majority of services, including military and security services associated with the activities of PMCs, the military requires to operate efficiently. They include, but are not exclusive to, Booz Allen Hamilton, Vinnell Corporation, DynCorp, BDM International Inc., and Science Application International Corporation (SAIC).

Booz Allen Hamilton was founded in 1914, starting out as an industry management consultancy. By 1915, it was undertaking business research, performing studies and statistical analyses. In 1940, the company helped prepare the US Navy for war. Then, in 1947, it got its first US Air Force contract, which led to contracts involving electronic intelligence. In the 1950s, the firm expanded into a number of areas including the electronics industry. By the mid-1950s, the company was awarded a contract to work on the Polaris nuclear submarine. It undertook its first overseas contract at this time. 1960 saw the company take the US Navy to the marketplace, paving the way for modern methods of ship acquisition. Throughout the 1970s and 1980s, the company was at the forefront of the revolution in technology and microelectronics. In 1990, it delivered critical systems capabilities in support of the allied forces during the Gulf War.[13]

Vinnell Corporation was founded in 1931. Originally a construction company, Vinnell went on to become experienced in managing military assignments during the Second World War. The company continued to expand in the post-Second World War era, taking advantage of the booming construction business in Asia. By the 1970s, the company had moved into the area of military training. They have now been training foreign forces for more than two decades, including the National Guard of Saudi Arabia. Today the company provides a broad spectrum of professional and technical services ranging from military training to educational and vocational training, and logistics support in the US and overseas.[14]

DynCorp was founded in 1946. The company is one of the largest employee-owned technology and services companies in the US, providing integrated technology, outsourcing, and technical solutions for public and private sectors worldwide. The company is employed by more than 40 federal agencies, including the DOD to supply aviation services, marine services, logistical support services, and personal and physical security services. As with the other companies, DynCorp operates globally.[15] The company was purchased by Computer Sciences Corporation in March 2003 for about US$950 million.

BDM International Inc. started business in 1960 as an information technology company. The company primarily operated in the defence sector before it diversified into information technology and computer systems. By 1997, 36 per cent of the company's revenue came from US defence work in information technology. In December 1997, the company was acquired by TRW. As with the other companies discussed here, BDM International Inc. has a wide customer base that includes many overseas clients.[16]

SAIC was founded in 1969. Again, as with the other companies, SAIC operates in the high technology industry; not surprising considering they started out as a research engineering company. The range of skills the company now offers to both the public and private sector is extensive and includes international maritime security solutions, military advice, and national security outsourcing. SAIC is a global corporation operating in 150 cities worldwide.[17]

All these companies have proved to be reliable and technically competent to supply a wide range of defence related services to the US government. A number of them have even specialised in particular defence contracts including developing a wide range of military programmes designed to train the US army in a range of combat and non-combat skills. By combining their knowledge of the military with their huge commercial resources, these companies have been able to apply unique solutions to difficult problems, even if they have not always been successful. However, the roles these companies are being considered for are changing to include military services which in the past would have been the sole domain of the military.

In the technical field, the US State Department in Columbia has hired PMCs, including DynCorp, California Microwave Systems, a unit of defence giant Northup Grumman, and Florida-based Airscan to supply more than 100 pilots and mechanics as part of a programme to eradicate Colombian coca and opium poppy fields.[18] The pilots fly a range of different aircrafts, including the planes that spray herbicides on the illicit crops, gunships that accompany the spraying mission, and the transport aircraft used to supply the operations. Others work as aviation mechanics, logistical experts and medical staff. At the same time, private defence contractors are also being hired to help Afghan forces eradicate poppy cultivation. The DOD is also considering using civilians to pilot unmanned aerial vehicles (UAVs) in military operations.[19] The introduction of civilian pilots is intended to speed up the development of UAVs by countering what many in the Office of the Secretary of Defense see as a pilot-rated community of Air Force Officers determined to slow development for fear of losing cockpits in which to fly.[20]

AirScan is another US company that has contracts with the Department of State, Department of the Army, US Air Force, the Angolan and Columbian governments, and numerous other customers to perform airborne surveillance and security missions as well as tactical support missions for the military. The company has seen a marked increase in its core business since 11 September, supplying affordable, tactical, real-time information to its customers to enhance their operational capability. The company's headquarters are in Rockledge, Florida. They currently operate a fleet of Cessna C-37 Skymasters outfitted with the latest electro/optical/infrared surveillance systems.[21]

As mentioned above, the US military also employ civilian specialists trained to operate, as well as maintain, complex battle systems including maintaining the B-2 stealth bomber. As weapons and equipment in the US military arsenal become more complex, the military is relying increasingly on teams of civilian contractors to ensure the serviceability of the equipment. Such individuals work alongside military personnel, sharing the same living space, as well as facing the same war zone dangers. It has been claimed that the war on terror is being fought with a higher concentration of civilian specialists than any other conflict. In Afghanistan, for example, civilians initially operated the Global Hawk, one of the US military's newest and most advanced systems, and which can only be maintained by those civilians who built it. As a practical matter therefore, as war fighting becomes increasingly automated with the use of computerised weapons systems becoming more widespread, the trend of employing private contractors will continue to grow for the foreseeable future.[22]

Another area of operations for PMCs is intelligence. This is an area PMCs and other private defence contractors are becoming increasingly involved in especially after 11 September. Companies in the US now provide a range of intelligence services to government agencies providing, for example, common user platforms for intelligence professionals that range from the Joint Task Force, Theater Joint Intelligence Centers, Theater Component Commands, Coalition Commands and Allied Partners to Service and Unit level Commands, the Department of Defence (DOD) and non-DOD national agencies. At the same time, companies are also providing programs for other members of the security communities including intelligence, law enforcement and homeland security agencies. Much of the work carried out by the private sector is in the area of technical support, particularly information technology services able to provide real-time intelligence, secure communications systems, force management, and airborne surveillance.[23] Companies also provide manpower to the intelligence community, the most obvious examples being the private contractors that worked at Abu Ghraib prison in Iraq.

CACI International, a Virginia-based defence contractor, and Titan Corporation, which has its headquarters in San Diego, California, provided civilian interrogators and interpreters to the US-run prison. According to job descriptions obtained by the Center for Public Integrity, CACI employees also performed a variety of other tasks in Iraq. Contractors undertook debriefing of personnel, intelligence report writing/quality control, and screening/interrogation of detainees at established holding centres. The contractors worked closely with military personnel, coordinating, for example, interrogations with military police units and military intelligence interrogation units assigned to support operations of the Theatre/ Division Interrogation Facility.[24] Middle to low level contractors were expected to operate in the military hierarchy, while senior level advisers worked directly with military commanders. Senior level advisers were also responsible for managing the work of subordinate contractors involved in interrogating detainees.[25]

A further area of operations for PMCs is running training programmes for the military as well as training foreign militaries on behalf of their government. The aim of these programmes is to enhance the ability of the militaries, particularly in developing countries, so they can conduct peacekeeping and humanitarian relief operations without direct US military support. The US ACRI programme has trained several thousand African soldiers in peacekeeping and crisis assistance tactics. ACOTA has now replaced this particular programme. The key difference between these programmes is that ACOTA tailors programmes to the needs of individual countries with emphasis on training trainers, not just troops.[26]

Moreover, by using PMCs on these programmes, the US government is also able to release military personnel to undertake other essential duties. At a time when the US military is being asked to perform more and more tasks with fewer and fewer personnel, outsourcing training programmes seems only sensible if the military are to meet the operational demands being placed on them. Neither does using a PMC necessarily mean lower standards. Many of MPRI's field staff held senior staff positions or were senior NCOs with over 20 years' experience in the US military.[27] Civilian staff can also enhance a programme by bringing continuity to it. Military personnel spend up to 18 months on a programme before being designated a new task. PMCs, on the other hand, can contract civilian instructors for the duration of the programme. Neither do these programmes appeal to career soldiers. Undertaking tasks outside of the army's mainstream career ladder for 18 months can have an impact on a soldier's promotional prospects, while using civilians removes this particular problem.[28]

Nor is it surprising that US PMCs are undertaking more and more nonessential duties that in the past would have been the responsibility of the US military, while the organisational structure of US PMCs places them in an ideal position to exploit this situation by mirroring a primary function, managing war, which the US military excels in. McGhie makes the same point when describing US PMCs,

> Noticeable about the US sector is the size of the companies involved. MPRI, Dyncorp, Armour Holdings, Vinnel, and Brown & Root each have large permanent staff and extensive databases of contract workers ... [Another] characteristic of US PMCs is their close co-operation with the US government, either working with or for the institution ... A third defining characteristic of US PMCs, with the exception of MPRI,[29] is that their core skill has always been logistic.[30]

MPRI is, according to Howe, the US PMC that exemplifies the growing government–company relationship.[31] This illustrates how the US government is becoming more and more involved with PMCs because of the possible advantages they offer over the US military in non-lethal functions, rather than combat capabilities. Such advantages range from fulfilment of national goals with financial and manpower savings, continuity of programmes by using the same staff, to supposedly less public scrutiny over their activities.[32] It is unlikely therefore, that US PMCs will disappear, but grow in strength, as the US military consider contracting out 214,000 military and civilian employee positions to the private sector over the next few years.[33] For the time being, the primary focus of US PMCs will remain non-core competencies, supporting functions that include logistics, intelligence, operating communications and weapons systems, maintenance, and running training programmes. This will allow the military to focus its energies on core functions such as combat operations.

Finally, PMCs do seem to have a more active role in US government policy. The US PMCs participation in America's war on drugs is an example, while this is likely to increase in the immediate future. US PMCs will invariably continue to supply pilots, mechanics, and other skilled personnel to the US State Department in Columbia in particular as part of a programme to eradicate Columbia's coca and poppy fields. Staff from PMCs also draw less attention to themselves, while the services they provide create a space for further involvement by the US government without putting the military on the ground. Thus, as long as the war on drugs continues, so the US government will continue to call on the services of US PMCs to help fight it.

UK government approach to outsourcing technical and non-core military services

The UK PMC market has not developed in the same way as the market in the US. This is mainly a reflection of different national attitudes held in each country towards the role of the market in contracting out military activities. Neither are the characteristics of the companies the same, particularly in terms of size and role. On the whole, UK PMCs are smaller than their American counterparts, while they have generally focused on the commercial market instead of government contracts normally favoured by US PMCs.[34] Even their historical roots are different. While UK PMCs can trace their roots to the mercenary outfits that operated in places such Africa during the 1960s and 1970s, US PMCs are rooted more or less in the engineering and service sector of the US economy. Even so, like their US cousins, UK PMCs are today professionally run commercial organisations, the official military transformed into the private sector as Spicer has put it.[35]

UK PMCs operate for a wide range of organisations, including both British and foreign governments, NGOs, and commercial organisations. A substantial amount of their business is with the corporate sector, providing a comprehensive array of security services or security related training. These services include, for example, country-specific risk analysis and supplying security consultants to advise governments and corporations on security strategies. Security is now such a crucial part of corporate planning when operating in and around war zones that the relationship which exists between these two groups is now much more embedded than in the past. UK PMCs are also less influenced by government contracts than PMCs in the US,[36] making them much more reliant on the commercial sector for business success. Support to government agencies, such as the MOD[37] and DFID, is relatively small when compared to the support US PMCs, such as MPRI, receive from the US government agencies. Even so, a relatively close relationship still exists between UK PMCs and government officials, but is much more informal and based around contacts established while still servicing in the military or working for government agencies, as a result of which the activities of UK PMCs are unlikely to compromise British interests according to Shearer.[38]

The organisational structure of UK PMCs reflects certain aspects of the armed forces. According to Westbury, UK PMCs tend to draw on a pool of Special Forces veterans,[39] which mean that they inherit the conservative nature, as well as other traits particular to Special Forces. Companies tend to be small, managed by only a handful of permanent staff, all of whom know each other and hold the details of their workforce on databases. The top companies are very careful about who works for them, preferring to hire colleagues whose record of accomplishment they know, or can verify. Former members of Special Forces or other elite units in the British Army appear to set up most UK PMCs, operating around networks of people who all know each other.[40] In this respect, the companies are far more homogenous than US PMCs. They are also, as a result, smaller and not normally part of large multinational corporations, as in the case of US PMCs.

Companies in the UK also use networks to track down former soldiers who have a particular military skill that a contract requires. Here, members of one network can easily link into another network. At the same time, members are likely to have served many years in the military, creating numerous contacts across the military before moving to the private sector. In this situation individuals tend to know who, or which unit, has the skills required for a particular job, or can find out through their network of contacts. As Kevin James, a former senior NCO with the Special Boat Service (SBS), explains: when transferring certain military skills associated with particular units to the private sector, it is normally undertaken by a retired member or members of that unit.[41] If the private sector, for example, wants an expert in maritime counter-terrorist operations they will invariably seek out retired members of the SBS, Britain's maritime counter terrorist experts. Explosive Ordinance Disposal (EOD) is another trade where private contractors wanting to undertake this type of work seek out retired members of 33 Regiment Royal Engineers, the British Army's bomb disposal experts.[42]

The fact that UK PMCs have close links with elite units, whose members would have served in the Falklands conflict, the Gulf War, and Northern Ireland, also suggests a reason why these companies are prepared to undertake particularly dangerous operations that use lethal force. These operations range from anti-piracy, close-quarter protection, and counter-terrorist activities, and invariably take place in some of the most dangerous parts of the world. Even so, as Beese explains, there appears to be a dividing line between those who served in the Falklands and appear willing to operate in a more offensive manner when conducting operations, and those whose military experience is with peacekeeping and tend therefore to emphasize the need to conduct operations in a more defensive manner.[43] Despite the different approaches, what is clear is that the high level of skills exhibited by UK PMCs has increased demand for their services. According to James, some recruits to Britain's Special Forces are joining with the sole intention of doing a minimum of three years before leaving to work in the more lucrative commercial sector.[44] Many former soldiers, not only those servicing in Special Forces, are also multi-skilled, as well as multilingual. As a result of this, individual employees can undertake more than one particular role during a contract, reducing the cost of the contract for the company while increasing the range of services offered. It also allows operations to remain relatively small.

As just noted, involvement with UK government agencies is not as widespread as in the US. This is in part due to the government's lack of understanding about what the industry represents and what services it can supply them with. Unlike US PMCs, which are seen as a valuable resource to be fully utilised by the government, UK companies are not viewed in the same way. This is because, undoubtedly, the result of their mercenary roots has placed a question mark over their legitimacy.

Consequently, the government has been hesitant to engage with them, preferring instead to maintain their distance. Unfortunately, instances such as the Sandline Affair in 1997 have not helped improve the situation, but instead exposed the very tensions in the relationship that have prevented the industry and government from working together. Ironically, while the government at present remains reluctant to engage with these companies, the same is not true in the US government. British companies have been involved in protecting US embassies in the Middle East and the US naval base in Bahrain.[45] More recently Aegis Defence Services was awarded a US$293 million contract by the Pentagon in May 2004 to act as the coordination and management focal point for all the reconstruction agencies working in Iraq. The company also provides protection for the US government's Project Management Office, and the Oil-For-Food Program.[46]

All this is not to say that some of these companies have never undertaken contracts for the UK government, but that such contracts are in areas far less controversial than the provision of military services. Armorgroup and Control Risks group have both worked for government agencies in the past and continue to do so today. From 1999 to 2001 ArmorGroup supplied mine action support immediately after the Kosovo conflict in response to a government appeal. The company worked in cooperation with the UN and other mine action organisations clearing mines and unexploded ordinance to facilitate the return of refugees to their homes. DFID's decision to fund commercial as well as not-for-profit and profit organisations to conduct humanitarian mine action projects after the Kosovo conflict came as a wake-up call for some in the NGO community who were quick to question whether commercial organisations could deliver sustainable national ownership. Control Risk Group has also supplied government agencies since 1982 with a range of services, including security advice.[47] More recently, both companies were awarded contracts by the FCO and DFID to supply private security for their officials working in Iraq. The contracts are worth about £25 million to the companies.[48] It is still not yet clear, however, whether the government intends to engage fully with these companies.

Whatever the final outcome, the government is already outsourcing technical and non-core military services. As with the US military, routine maintenance of military equipment is now being carried out by private contractors. An example of this is warship maintenance which is now undertaken by engineering companies such as Babcock Engineering Services. The same company maintains C-130 Hercules and Hawk aircraft as well as providing services to the army's Armour Centre maintaining the entire range of the army's tracked vehicles and delivering driver, gunnery and communication training.[49] The range of work the company engages in also demonstrates the level of penetration of the military by engineering firms. Engineering, however, is not the only industry that is benefiting from outsourcing. The MOD has considerable and long-standing experience of contracting out the provision of services. Furthermore, outsourcing remains a key method of achieving targets set under the Public Service Agreement, while contracts are often in the order of five to seven years in length.[50] This aside, however, there appears little space at present for UK PMCs inside the military machine, while the obvious reason for excluding them is the controversial nature of the services they supply. The idea of employing private contractors to protect soldiers and government officials appears absurd, but that is exactly what is happening in Iraq, where the US government is using private contractors and not soldiers to guard the Green Zone. US contractors were even contracted to protect Paul Bremer, the US overlord to Iraq.[51] Providing such services though is only a small part of their business. For UK PMCs which already provide less controversial services to government agencies it is disappointing that the UK government still appears reluctant to outsource more controversial services to them because of the worries about the consequences if it goes wrong.

The outsourcing of military roles in Iraq

So long as wars continue so the demand for military expertise will exist, while PMCs will carry on taking advantage of any failure on the part of governments to meet these demands. Such failure will see PMCs continue to play a significant part in international security for the foreseeable future. They will do so for all kinds of clients so long as the reasons responsible for the expansion of the industry remain. At present, there appear no other social forces on the horizon to dampen this expansion. Moreover, the very forces driving the industry do not appear to be slowing. This means the demand for military and security services and the ability of companies to provide them will continue to grow at an ever increasing rate. But, while the future looks promising for PMCs there are still significant problems facing the industry and governments that both would be well advised to confront now rather than later.

Iraq has shown us the extent to which some governments are now willing to go in outsourcing military services to PMCs. Nor has the industry been slow to take up the challenge. And while the actions of some of the companies need to be questioned, they have on the whole performed satisfactorily and in specific cases heroically. PMCs have clearly demonstrated their utility to the US government in the war in Iraq. But while the war has shown that there are advantages to using them, there are also the pitfalls for governments employing their services. Neither is it clear whether the advantages always outweigh the disadvantages. A decision to use PMCs is usually based on a judgement, while at the same time officials are normally responding to a set of external pressures that require immediate action and as a result of which they do get it wrong, and even when they get it right their judgement is sometimes called into question. The decision by the Pentagon in May 2004 to coordinate and manage all PSC work in Iraq through a single point of contact was seen by many of the security companies in the country as a sensible move. The choice of Aegis Defence Services as the coordinating and management agent, however, raised numerous problems for Pentagon officials because of the history of Aegis's Chief Executive, Tim Spicer.[52] Nor were the other PSCs in Iraq particularly happy with the decision to award such a major contract to Aegis, since the company was new with very little experience of the security business when it won the contract.[53] Getting it right every time is not going to happen; officials will make mistakes as they try to understand the nature of this new actor and how and what it should be employed to do for them. But that aside, there are key indicators as to the way the market for their services may develop.

Hiring private contractors whether local or ex-patriates in Iraq has reduced some of the political risk associated with deploying US troops. But these are only two, though very important, wars the US are involved in. There are numerous other smaller wars, particularly in places such as Africa, where US involvement is very limited because the deployment of troops is fraught with political risk. Nor is this a solely US problem. The body-bag syndrome is now impacting on other militaries, and not just those in the West. Outsourcing to the private sector to get around this particular problem is one answer, but the approach is also laden with potential pitfalls. Neither should political expediency marginalise these problems. Appealing as it may sound, allowing contractors to take over from soldiers will present PMCs with huge challenges that could see them fail. So, while it may appear the politically safer option, it may not be in everyone's interests, including politicians and those working for PMCs, to outsource too much responsibility to companies. After all, no one wants to see troops being sent in to rescue a PMC because they failed to gauge correctly the seriousness of the situation they were getting themselves involved in and, as a result, found they had neither the manpower nor resources to cope.

One of the most worrying trends facing governments, and made worse by the war, is the poaching of soldiers by PMCs to fill their corporate ranks. While many of the established companies refuse to employ young soldiers, many of whom will have asked for Premature Voluntary Release (PVR) to cash in on the lucrative job opportunities the war offers, not all companies are so obliging. Iraq has turned into a numbers game for the companies. Some of the companies have even resorted to employing bouncers and security guards, raising new concerns about the calibre of staff providing vital protection services. So worrying has the problem become for MOD officials that in April 2004 they approached the companies working in Iraq and asked them not to recruit from among serving soldiers.[54] But it is not clear what impact, if any, the request had. Fiji is facing the same problem. The military, however, are finding ways to counter the problem. Some soldiers who have worked in Iraq for PMCs have even been reapplying to join the military once their contract is up.[55] The extent of the problem is not confined to the UK and Fiji. This is becoming a global problem that is set to get worse as the industry expands. What is needed is a global initiative that will force the industry to face up to its responsibility and not place future stress on over-stretched military forces by recruiting their soldiers. If such an approach is not possible then the only other option is for individual governments to take action to prevent a brawn drain from their militaries.

A further issue is the possibility of PMCs failing to fulfil their contractual obligations. Admittedly, this is not yet a major problem, even though US troops experienced equipment deficiencies in Iraq as a result of contractors not delivering the equipment when required. There is a real potential, however, for the problem to get much worse. According to David Wood who was a reporter with the invasion force, 'US troops in Iraq suffered months of unnecessarily poor living conditions because some civilian contractors hired by the Army for logistical support failed to show up'.[56] Civilian contractors may perform well in peacetime, but performing in the middle of a war is another matter. According to Lt General Charles Mahan, 'we thought we could depend on the industry to perform these kinds of functions'.[57] Civilian contractors cannot be ordered into a war zone. In this respect, it does not matter that they have signed a contract since they can always terminate it if conditions are too dangerous, leaving the military to pick up the pieces.

It is unfair to suggest, however, that contractors are not prepared to take the type of risks that are routinely asked of soldiers. The death toll among contractors in Iraq is not insignificant and the numbers are increasing all the time. Neither are the companies pulling out because of it. The majority of PMCs remain adamant that they will remain while their services are required. But asking a civilian who has combat experience to get in harm's way is very different to asking someone with no experience of war to do the same. Many of the service companies rely on individuals with no military experience to drive trucks, serve meals, or maintain the camps US soldiers live in. In future, it may be much harder for these companies to convince such individuals that their skills are valued and that they should continue to work in such dangerous environments.

The role of PMCs in Iraq also raises the issue of control. Even the companies recognise the need for regulation. Unfortunately, the only governments that have introduced legislation are South Africa, with the Foreign Military Assistance Act and the US with International Traffic in Arms Regulation. As Chapter 7 explains, both these attempts have on occasions failed to stop the most determined contractors from breaking the law. More importantly, governments need to do more to ensure the industry is stringently regulated. At the same time they also need to embed within the companies and their employees principles of human rights so that their actions conform to International Humanitarian Law (IHL). Furthermore, the International Committee of the Red Cross (ICRC) will hold to account any government that allows a PMC operating from its territory to disregard these principles.

The war on terror

Following the attacks on 11 September, the economic outlook for the industry has never looked brighter. Although the global economy then faced an economic downturn, the share prices of some of the leading PMCs jumped roughly by 50 per cent with some companies even seeing their share price more than double.[58] The rise in share prices after the attack had, according to one journalist, levied the equivalent of a 'security tax' on the global economy.[59] The attacks placed security at the top of every business leader's agenda increasing the demand for security and which PMCs could only benefit from. The need to protect company's staff, property and assets suddenly took on a new urgency as businesses reassessed the risks of operating in a global marketplace which left them vulnerable to attacks.

These concerns were further heightened in the aftermath of the bomb blasts on the British Consulate and the HSBC Bank in Istanbul which killed at least 27 people and wounded nearly 450.[60] It is fair to say that the events of 11 September and the bombs in Istanbul, which have brought so much suffering to so many people, are also responsible for giving the security industry a new lease of life.

The response from governments and commercial organisations to these attacks, and the threat of more terrorist attacks to come, has helped to consolidate and improve the industry's health. As the spokesman for the Pentagon's Defence Security Cooperation Agency explained, 'the war on terrorism is the full employment act for these guys. A lot of people have said, Ding ding ding, gravy train'.[61]

Their role in the war on terrorism is truly global. In Afghanistan security contractors are involved with developing security programmes and providing security management and armed security services to support the country's reconstruction projects. They are also helping to train the new Afghan police force as well as providing security for the country's president. Even the US operation to eradicate the Afghan opium crop will involve the use of private defence contractors. Companies also provide logistics to units serving in the country, while Brown & Root have been responsible for building and operating military bases all over central Asia.[62] Afghanistan and Iraq are not the only countries receiving military assistance to defeat terrorism. The US military is involved in giving military assistance globally. And as that assistance increases, PMCs will be expected to play their part. One of their primary tasks will likely be retraining state militaries and local enforcement agencies, running the type of military training programmes MPRI ran in the Balkans in the mid-1990s. Finally, the war against terrorism has seen the US and its allies operating in new regions that have taken on a greater strategic urgency. Central Asia and Yemen are two such areas. Such deployments come at a cost. Troop levels may need to be reduced in other areas for these deployments to go ahead. This will invariably create a gap that PMCs will want to fill.

Conclusion

The long-term future for the private military/security industry has never looked better. The fall in the number of young people enlisting in the military, technology, and the shift in the structure of modern society have all played a part in making space for PMCs to operate. Reduction in support for military intervention has also created opportunities for PMCs to take on non-core military functions. In respect to this last point, we may be witnessing the start of a process that will eventually lead to a division of labour within the military itself. It is not difficult to envisage a military where the role of the soldier is simply to fight or, when not fighting, training to fight, while all non-core military functions are outsourced to contractors. Fighting is, after all, the primary function of a military, whereas the other services are simply there to support them. What we do not know yet is whether the process will succeed.

As mentioned above, approaches to outsourcing also differ between governments. The roles of PMCs tend to reflect the functions that their state militaries do extremely well in. At the same time they remain tightly associated with their national context. US PMCs illustrate this perfectly. Companies such as MPRI are simply an extension of the military machine, carrying out whatever duty the government assigns it. They may be a commercial organisation, but in all other areas they very closely resemble the military. In the UK the situation is different. The qualities of the companies reflect the qualities frequently found in Special Forces. This is not surprising considering many members of Special Forces find their way into the ranks of such companies and as a result of which they tend to maintain a low profile when operating.

So far our primary concern has been with outsourcing non-core military functions to PMCs. A process that is unlikely to slow down soon. And state militaries are not the only organisations to benefit from the services offered by PMCs. Other government departments and organisations are also willing recipients of their services. As the next chapter discusses, PMCs are now supplying security services to an array of customers from development agencies, international organisations, and multinational corporations to NGOs. The majority of these services are security related, supplying services and advice to organisations working in dangerous environments around the world. That said, in the UK in particular, a different market is opening up to these companies. New approaches to SSR, which was once the sole responsibility of government agencies, are involving more and more private sector actors. And while such actors tend to be large management consultancy firms, this is starting to change, as is already happening in the US with PMCs taking on more and more SSR roles.

6

THE ROLE OF PMCs IN
A CHANGING GLOBAL
ENVIRONMENT

Introduction

The new post-Cold War environment continues to be plagued by war. This is especially so in Africa, where internal and regional conflicts are a regular occurrence in many parts of the Continent. These conflicts have come about as a result of the end of the Cold War and the patronage politics associated with it. No longer are political elites in Africa able to look to the superpowers for military support when threatened with civil war. In the past, either the US or the Soviet Union would have helped to suppress the very internal tensions that today fuel many of the conflicts the Continent is now experiencing. According to Kaldor, such conflicts have replaced the relative stability of the Cold War and are characterised by a multiplicity of types of fighting units ranging from public and private, state and non-state, to a combination of both types.[1] At the same time, the socio/political and economic environment of these wars is dominated by warlord politics, shadow economies, and strategic complexes.

The role of private security in this environment has been widespread and controversial. PMCs have been active in many parts of the Third World, but particularly Africa. They have taken part in the civil wars in Angola, Sierra Leone, and the Democratic Republic of Congo. PMCs have fought for governments, as in the case of Angola and Sierra Leone, but have also provided assistance and support to rebel groups. Johan Niemöllar, a former CCB operative, recruited former South African Soldiers to fight for UNITA. At the same time, he was employed by former Mobutist General Kpama Baramoto to establish arms caches in Central Africa and hire forces to support a move against Kabila.[2] Payment for such support may well have derived from profits drawn from activities more closely associated with the shadow economy of these countries, rather than through normal economic activities. And while they may or may not take a direct part in the shadow economy, PMCs can benefit from it through payment for the security services they supply. As controversial as their participation in civil wars, is their growing participation in emerging strategic complexes. Such complexes include international NGO, government, military establishments, international financial institutions, international organisations, the business sector, and so on. They are strategic, in the sense of pursuing a radical agenda of social transformation in the interest of global stability.[3] To achieve this, there has occurred an expansion of previous complexes that in the past were represented by the development or aid sector. Today such complexes constitute a network of strategic governance relations that are increasingly privatised and militarised.[4] Moreover, they now include PMCs and PSCs.

The chapter first sets out to explain the political nature of warlordism and its reliance on the shadow economy which now dominates many parts of the African economy. Rulers adopt the political logic of warlordism instead of strengthening state institutions to marginalise the threats presented by opposing strongmen. At the same time, they have had to contend with, but also rely on, an array of armed groups including PMCs to secure their powerbase. Following on from this, the chapter then sets out to discuss the merger of development and security into new strategic complexes. These complexes include an array of actors from the public/ private, military/civilian sectors, while the purpose of these actors is to resolve political impasses, undertake development, and improve on the lack of security that characterises most new wars. Finally, the chapter explains the function of private security within this new arena. Here particular attention is given to their role in managing security sector reform programmes, working for multinational corporations and international organisations, and giving quiet support to NGOs, while also examining the dynamics played out in each of these roles.

Warlordism and the shadow economy

Warlordism is a political and economic system that represents an alternative system of power, profit, and new forms of legitimacy. The structure of warlordism has evolved distinct strategies for political, economic and cultural survival that offer a very different approach from the institutionalisation of sovereign authority recognised in the West.[5] Such strategies include the use of shadow economies to promote economic interests, the use of marginalised youth alienated from the old patrimonial network, to utilising interlinking networks of social, political and commercial interests to retain power, instead of using formal structures of state bureaucracy. Furthermore, in order to strengthen their position, rulers take the politically easy option of taking control of strategic resources with the help of external support willing to risk investing in enclave operations, instead of supporting tax-generating groups. Acquiring funds from private sources, and not the general population, allows rulers to place further demands on the population for resources, even if this drives down general productivity.[6] At the same time, by safeguarding resources which would normally pay for public services, rulers are able to pay to keep the loyalty of warlords and strongmen.

Like patrimonialism, clientalism is central to warlordism. The reassertion of local politics has taken a variety of forms, including new or renewed forms of local and international clientalism. For the rulers of weak states, clientalism is now about forming links to different local and international groups to enhance one's security, instead of relying on an outside power. Rulers, for example, have entered into agreement with local civilian militias, such as the Kamajohs in

Sierra Leone, for support against rival strongmen. Rulers have also entered into international agreement with multinational corporations, with their vast array of resources, to protect them against threats from rival strongmen. Furthermore, the services provided by such groups have empowered rulers, while also making them subordinate to the demands of such groups.

Warlordism and clientalism developed out of the patronage client politics of the Cold War. However, while there are similarities between warlordism and patronage client politics to do with the structure and strategies, there are significant differences in the way the international environment has influenced how each works. The patron/client relationship was defined by Cold War politics, as where warlordism/client relationship is defined by regime survival. Consequently, rulers of weak states have had to adapt the structures and systems developed during the Cold War to a new set of priorities directly relating to regime survival. In this respect, the financial loans and development assistance given during the Cold War provided the basis for expanding the bureaucracies of developmental states, while such bureaucracies were utilised, if only in name, to receive external aid, and by rulers to build alliances and forge clientage networks.[7] Moreover, such bureaucracies were the main conduits for external assistance.

While the process was never perfect, the end of the Cold War, and structural adjustment programmes, has further increased the pace of debureaucratisation in weak states. Developmental aid from international organisations, including the World Bank, the International Monetary Fund (IMF), and donor governments, is now given on the understanding that state bureaucracy will be cut even further. Undertaking such reforms also shows how serious rulers are to other external audiences. This in turn has helped rulers to establish new patronage/client networks using the country's natural resources and the shadow economy to entice new partners.[8]

The shadow economy

Shadow economies are the consequence of the inability of rulers of weak states to control the economic activities that go on inside their territorial borders. The private, non-bureaucratic arrangements that are an essential part of the shadow economies also pose a real threat to the authority of incumbent rulers of weak states. This is because they deny them revenue to spend on public amenities, such as education, thus further alienating the public from the state. In certain countries, such as Liberia for example during the 1990s, the shadow economy had more or less replaced the official economy. No longer was the Liberian state at the centre of the market. Instead, Taylor occupied that place.[9] By privatising all economic activity inside the territorial areas he controlled, he could grant selective privileges or impose punishment on traders operating outside the unofficial boundaries he set up. Furthermore, the shift from a state regulated system to a private personal syndicate system of control over commerce, abjures all remaining state responsibility from the state to the public, thus threatening the very existence of the state.

Participants of shadow economies benefit from the lack of stability caused by new wars, because they can charge higher prices for commodities, and thus increase their profits. Duffield describes shadow economies as essentially extra-legal and evasive in their mode of operation. Continuing, he argues that participants in the shadow economy frequently operate outside formal or state regulated systems, operating instead through complex relations of collusion or complicity while in competition with the state regulated systems.[10] Such a description captures the true nature of the activity; unrecorded, unregulated, unconventional, extra-legal, and cross-border in character.

Globalisation has enabled the shadow economy to expand at an incredible rate and the introduction of market deregulation and privatisation has opened up opportunities for foreign companies exhibiting different levels of perceived legitimacy. Such foreign firms have established themselves on both sides of the supply network, fuelling new wars or attempting to prevent them by denying access to resources. For example, the international environmental watchdog, Global Witness, has appealed to the Danish company, DLH Nordisk, to stop buying conflict timber.[11] DLH Nordisk was amongst a number of European timber buyers who purchased timber from Liberian logging companies, several of which were directly involved in prolonging the conflict in West Africa.

Finally, the shadow economy is crucial to any ideas rulers may have about adopting warlord politics. Rulers who choose to address internal threats from rival strongmen using warlord politics must first transform their political authority into effective means of controlling strategic resources and other revenue-generating business. Then, instead of relying on formal state institutions, they are able to purchase weapons or loyalty to protect themselves from internal threats to their position by using the shadow economy to exploit such resources, as the following section discusses.

How rulers adopt warlord politics

As discussed above, the end of the Cold War changed the nature of opportunities available to rulers of the developing world, who could no longer rely on previous strategies of governing as they attempted to respond to changes to the world economy. Again, such changes had been played out amid major changes to the economies of the developing world. This has upset old estimates of benefits while at the same time leaving rulers to decide what strategy to adopt to remain in power. Should rulers respond by cultivating popular legitimacy, complete the privatisation of state power, or adopt intermediate strategies? This is where the problem lies. Rulers whose authority is fractured face the stark reality of either turning to the electorate, who may demand more than just democratic representation, but also socio-economic rights that the ruler may not be able to provide,[12] or finding an alternative strategy for holding on to power. For some, warlord politics offered such an alternative. It allowed rulers the opportunity to enter into partnerships with foreign firms. This arrangement allowed both to profit from commercial opportunities created by disorder and the fractured authority of the state, while at the same time marginalising threatening strongmen.

The threat posed by strongmen is, in part, the consequence of disregarding the pursuit of a broader project of creating independent state institutions capable of serving a general interest distinct from the interests of the ruler. Economic investment is ignored if it looks as if political opponents will benefit, while other state responsibilities are contracted out to foreign firms, or loyal strongmen, again denying economic opportunities to political rivals. The lack of bureaucratic institutions also means these two groups take on a wider range of state responsibilities, improving their economic position even further, in return for remaining loyal to the ruler. This is particularly so for foreign firms reliant on the wishes of the ruler to operate their businesses. As such, they pose a minimal or zero threat. Thus, by generating wealth from hiving off state responsibilities, which is converted into political authority, rulers are able to strengthen their position by purchasing further loyalty and more weapons. Rulers, in effect, jettison all pretences of serving the interests of a public harbouring dangerous rivals or unruly citizens.[13]

The critical distinction between warlord politics and the state, even weak states, is the triumph of private over collective interests.[14] For those political leaders who espouse the former, informal shadow networks, represented by informal commercial networks of business interests operating outside the state regulated system, come to take the place of state bureaucracies as the means of staying in power. Taken to its logical conclusion, warlord politics sees collective and private authorities resembling one another, but with the emphasis on different values. For example, each type of authority provides security. But in the case of collective authority, security is a right of membership to the state, while in the private sphere security is a consequence of other factors, most notably economic interests.

Rulers who adopt warlord politics do not necessarily find their position improved and they can still face threats from strongmen excluded from any political dialogue designed to accommodate opponents. Rival strongmen who possess the necessary military power to keep rulers at bay are also able to control areas of territories where strategic resources are located. Then, by establishing a rival network of alliances linked to the international economy, they can exploit such resources to advance their political ambitions. As Duffield points out, today's warlord needs to act locally but think globally to be successful.[15]

Charles Taylor's National Patriotic Front of Liberia (NPFL), for example, was particularly successful at using foreign firms to pioneer links with the global economy to help him establish authority in Liberia. From its early days in 1989, the NPFL exercised control over much of Liberia, as well as parts of Sierra Leone involved in the mining of diamonds. The NPFL soon dispensed with the formal structures of state, establishing their own banking system, TV station, even introducing their own currency.

Foreign firms were crucial in consolidating Taylor's power over areas that he controlled. In order to strengthen his authority, Taylor established a vigorous trade in timber, diamonds, and agricultural products with other countries through foreign firms and commercial networks.[16] Taylor was soon working with the Firestone Tire and Rubber Company, whose links to the country go back as far as 1926. An agreement signed by both sides saw Taylor's security forces used to control workers on the plantation, allowing Firestone to concentrate on selling rubber abroad. In return for protecting Firestone's plantations, Taylor's regime was provided with communication facilities and a supply base that helped facilitate the NPFL's attack on ECOMOG's[17] position in Monrovia in October 1992. While Firestone's investment played no direct part in the attack, Taylor was able to release funds from other places by using Firestone's investment to cover the gap that this would have made in his financial resources.[18] Rubber was not the only raw material Taylor was able to exploit. By the early 1990s, Taylor had become France's third largest supplier of tropical hardwoods.[19] Again, he was able to do this by using his forces to control areas where the material was located. As a result, foreign companies became an important source of hard currency, while denying the same source to opponents. By only relying on foreign firms and intermediaries, compared to dependency on bureaucracies, Taylor had developed an innovative way of integrating himself into the world economy, while maintaining fractured authority in the country.

By adopting the same approach of acquiring resources by controlling commercial networks, Taylor was able to fund his armed struggle against President Samuel Doe, who had done exactly the same during his 10 years in power.[20] In the end, however, Taylor was able to seize control of most of Liberian commercial opportunities while marginalising the need to control specific territories. Doe's dismantling of all state institutions meant that, like him, Taylor would have to lock himself into a system of political/commercial networks from close contacts, rather than state institutions, if he was to raise the funds he needed to sustain his military force. Without his military force, Taylor himself would become the next target of strongmen. For Taylor, breaking the cycle of violence created by warlord politics was not an option, he had to maintain it or be ousted. Finally, the closure of commercial networks, by preventing access by strongmen to strategic resources and other revenue-generating means, is crucial to any ruler attempting to break the cycle of violence warlord politics creates, while at the same time allowing the introduction of new state institutions to regulate commercial transactions in the country. Yet the cycle has been broken in Angola and Sierra Leone, if only for a very short period by using PMCs as is discussed later in the chapter.

Factionalism

As with warlord politics, rulers face the same problem of breaking cycles of violence that exist with factionalism. Disenfranchised groups have also sought to challenge the ruler's legitimacy, causing fractures in the ruler's authority. Again, these groups are typically the product of new wars, which has resulted in the dispersal of the ruler's authority to other interested parties. The following section, however, is only concerned with groups that have been able to threaten the ruler's authority in the same way as strongmen, and thus forced rulers to commercialise and privatise politics and security arrangements to stay in power. For the purpose of analysis, a multiplicity of types of fighting units is included along with paramilitary groups, self-defence units, foreign armies, and mercenaries.

The term 'paramilitary group' covers an array of armed organisations willing to give or deny their loyalty to a ruler or regime. Often established by rulers to distance them from the more extreme forms of violence, this has not always ensured their total loyalty. Paramilitary groups tend towards securing their own interests, and, while they may be loyal to one particular ruler, they are just as capable of denying that same ruler loyalty if they believe their interests are not being served. Thus, paramilitary groups, such as the notorious West Side Boys in Sierra Leone,[21] having seen their support marginalised because of the decaying patrimonial networks, chose to resort to violence as an alternative method of getting what they wanted from the government, or to control resources of their own. Such groups are normally autonomous and generally centred around an individual leader. They are mostly composed of redundant soldiers, or whole units of redundant soldiers, while they may also include criminals and child soldiers.

Other groups involved in private violence may also on occasion represent a threat to the ruler's authority, although this is not always the case, particularly when such groups are no more than self-defence groups. Normally composed of volunteers, the intention of these groups is to defend the citizens of their locality from physical attack. They are a consequence of the central authority's inability to carry out one of its most fundamental responsibilities, that of ensuring the protection of its own citizens. Certainly, local agreements between rulers and those who command self-defence groups may be necessary to ensure the authority of the ruler is not threatened. This is especially so when self-defence groups extend their responsibilities to include roles normally associated with state militaries. This was exactly the position of the Kamajohs in Sierra Leone who, during the country's civil war, emerged as a new military force.[22] As Hirsch points out, the rural villages had traditionally turned to the Kamajohs for protection, so it should not have come as a surprise when their leader, Chief Hinga Norman,[23] mobilised them to help defeat the RUF, rather than trust the army, in whom he had little confidence, to accomplish the task.[24]

Breaking the cycle of violence perpetrated by paramilitary groups is again essential in restoring the ruler's authority as it is with strongmen violence. The emergence of self-defence groups, on the other hand, has come about because of the lack of protection offered by the state against this violence. Even though it is difficult to stop these groups, they resort to violence only to survive. The restoration of state institutions in place of warlord politics would significantly alter the situation. Since the aim of these groups is not the seizure of power but simply survival, the redistribution of state resources to include them would remove their need to turn to violence. To achieve this, security sector reform is essential.

Foreign armies

The need for rulers, strongmen, or factions to continue to rely on violence because of the fractured political authority of the state is also aggravated by the military support given by foreign governments to such groups. There are, of course, numerous reasons why such support is forthcoming, though ultimately it reflects a concern for national security, the political situation, or economic advantage in the country offering the support. Military support for those engaged in civil wars is also far more forthcoming from countries bordering civil wars, though this is not always the case. The Mobutu regime, for example, was actively involved in three neighbours' wars. Mobutu supported the Khartoum government in its war against the Southern Sudanese rebellion; the government allowed its territory to serve as a rear base for attacks by armed movements against Uganda, Rwanda, and Burundi; and Mobutu gave support to UNITA. Thus, there was formidable regional support for Kabila's rebellion,[25] while five immediate neighbours of what is now DRC took advantage of the country's fractured authority to further their security needs by actively contributing to the overthrow of Mobutu's regime.[26]

At the same time, there is little evidence to suggest any of the warring sides relied on PMCs for combat or combat support operations.[27] And while PMCs were not slow in offering their services to Mobutu, they appear not to have been as successful as they might have expected to be.[28] This is not to suggest PMCs were in no way involved in the country at this time. A number of PMCs did secure deals with Kabila's government, while other PMCs were prevented from participating in the conflict by their own government or domestic opposition.[29] Furthermore, because Kabila's government was able to gain the support from foreign armies, those PMCs present found their ability to influence the outcome of the war was marginalised, if not removed. Instead, the conflict seems to have attracted elements of the classic mercenary network, while PSCs did provide an array of other services, including protecting mines, supplying arms, and offering training.[30]

Where PMCs have worked for rulers, they have often managed to reduce the level of violence confronting those rulers, while in two countries they dramatically reduced the level of violence associated with warlord politics. Angola and Sierra Leone are prime examples of countries where the political elite frequently engaged in warlord politics. In both countries, EO was able to use its superior training and skills to stop the fighting and thus break the cycle of violence that had plagued the country for nearly 20 years, if only for a short period. Contrary to the civil wars in Angola and Sierra Leone, the civil war in the DRC involved five foreign armies, none of which has been able to stop the fighting, while at least one has used commercial connections to take advantage of the country's natural resources.[31]

The dynamics of new wars and emerging structures of conflict

Invariably, new wars are, in part, the consequence of warlord politics that has led to the struggle to control strategic resources and thus control over shadow networks. The dynamics of new wars are not as straightforward as they might first appear. The following section explains what is meant by new wars and then provides an explanation of the dynamics that drive new wars. While Keen,[32] sees economic calculations driving new wars, for Kaldor, new wars are about identity politics, that is, the claim to power on the basis of a particular identity and not geopolitical or ideological goals associated with the Cold War.[33] In this respect, they differ from earlier wars in terms of their goals, the method of warfare, and the way they are financed. The strategies of new wars include, but are not inclusive of, guerrilla warfare, counter-insurgency, terrorism, and genocide.[34] What all new wars have in common, however, is their reliance on the global economy for finance, making the war economy of new wars different to earlier war economies.

Unlike conventional wars between states or even civil wars, new wars, as with violent peace, tend to blur the distinction between war and peace, as opposed to absolute or opposed conditions.[35] According to Duffield, countries that experience new wars are characterised by high levels of unemployment and underemployment, thus exhibiting debureaucratised and fragmented public authority combined with a high level of political autonomy among political elites in the country.[36] For these two social groups, as well as other disenfranchised groups, new wars represent access to different forms of protection, legitimacy, and rights to wealth, which has to do with trans-border shadow trade. Warring factions lock their local network into global networks. This in turn gives them the means to realise their assets while also making provisions for the future. For this to occur, however, sufficient power is required by a local warlord or strongman to mobilise a local network to the extent necessary to engage at the global level, but this also increases violence in the area of strategic resources and rent seeking opportunities as the need to maintain links into the global economy increases.

New wars also fail to draw distinctions, in the traditional sense, between civilian and combatants. Within new wars, individuals come to identify with a particular ethnic group, which at the extreme margins of violence have witnessed attempts to wipe out such groups through genocide. At the same time, the new war economies are decentralised, attracting few participants, apart from the many unemployed youths common in such economies, while combatants are heavily dependent on strategic resources, including diamonds and tropical woods, to finance their fighting. And where this is not possible, they have also resorted to extortion, plunder, or external assistance. New wars are all about taking over networks surrounding strategic resources or other revenue-generating projects, with little if any consideration given to the means employed.

State militaries have also turned to these methods of fighting and financing new wars. Political and security problems have forced rulers to take extreme measures to protect themselves against rival ethnic groups, while the domestic production necessary to finance new wars is either non-existent or in rapid decline. This is because of the inability of the state to compete against global competition as a result of the destruction or interruption to their economy caused by the war. In the end, new war economies are characterised by poverty, communication breakdown, resources competition, social exclusion, and criminality, which, according to Duffield, lead to various forms of collapse, chaos and regression for states engaged in new wars.[37]

In the end, turning politics into business is not about increasing the role of commercial calculations to do with private violence or foreign firms determining the agendas of developmental states. It is more complex and goes to the core of the paradox of globalisation.[38] The developmental state also channelled externally-provided aid for similar purposes during the Cold War. In this respect, there is nothing new about Reno's claim to do with private security companies helping weak states to manage and control warlords and strongman entities.[39] In each case, foreign actors helped shape internal outcomes. What is new in the post-Cold War era is that the rulers of weak states are no longer dependent on the resources supplied by Western governments to reward those that pose a threat to their authority, and thus reduce the influence Western governments have over them. Instead, rulers are now able to forge links to the international economy through multinational companies and their link to the private security industry. This has given rulers access to wealth-generating opportunities, while also creating new forms of legitimacy and conflict management. Furthermore, the loss of control over the international economy through deregulation has both fuelled new wars and made them private affairs free from outside interference. All this has of course fed into the new development/security arena, and the danger involved in underdevelopment. As Duffield notes, it is not so much the underdevelopment that is the problem, but the ability of rulers of weak states to exploit the advantages that globalisation has given them.[40]

Such involvement in shadow networks compromises efforts to introduce good government based around development and security aims that are themselves based in part on market openness. A similar problem exists for rulers of weak states that enter into contractual relationships with foreign firms. Here, the responsibility of the international community is to ensure market openness enhances, rather than compromises, good government. This is because difficulties, setbacks, or even conflict itself, all seen as a hindrance to market openness that can restrict and impede progress, can be avoided if not eliminated.[41] Thus, the business sector by playing a greater role in helping to develop free markets, including those companies providing private military and security services, can promote greater development and security that should result in a more enduring peace.

The difficulties for rulers of weak states trying to promote good government is that deregulation and liberalisation of the market has encouraged an increase in the creation of wealth and a new social order, based on neoliberal ideas, to take advantage of that wealth. This has all been done as a consequence of promoting support for the informalisation of the economy, parallel trade, illegal international transactions, and the proliferation of new wars. Neither is it possible to separate both sets of characters. As Reno points out, there is a contradiction within globalisation in that, as new wars increasingly disrupt the international economy, the more the international community will call for openness and fewer restrictions.[42] Consequently, we are likely to experience a marked increase in all types of private protection as a result of such disruption. Even so, while those with plenty of wealth will be able to remove or simply protect themselves from the contradiction, others will be less fortunate. In this situation, many may come to rely on protection offered by their employers able to pay for the private sector to undertake the responsibility, or simply go without any type of protection. In the case of the government, private military help may be necessary to establish a safe environment from which the general population can benefit.

Tackling new wars is problematic. As the next section explains, the global economy has helped to fuel new wars by allowing strategic resources to flow out of Africa, and for weapons to flow in, making the socio/political problems even more intractable. In responding to the crisis, the international community is transforming the social systems of developing countries through multilevel and increasingly non-territorial decision-making networks.[43] These networks, what Duffield calls strategic complexes, bring together governments, international agencies, and the private sector in new and imaginative ways. In doing so, the anticipation is they will be able to transform the social and political environment to create an endurable peace in countries plagued by new wars.

The link between new wars and the global economy

Globalisation represents a number of processes that are working to generate increased interconnection and interdependence between states and societies.[44] Globalisation, according to Clark, denotes intensity in the movement of international interaction.[45] The nature of that interaction, however, is not always peaceful. New wars and shadow economies, based around the flow of capital, are a distinct feature of this interaction in many parts of Africa, linking resource-laden regions to the global economy. Here, interaction refers to that between international financial and economic systems with a range of actors, including political leaders, warlords, multinational corporations, and international organisations all with a political agenda. In this environment of potential hostility, PMCs play a crucial role protecting the transfer of capital flows for their clients. We therefore need to examine PMCs in the context of transnational capital, while also examining their relationship to multinational corporations because of the impact this group has on assisting capital flow, a role that is becoming increasingly important to the world economy.

Strategic complexes

Strategic complexes consist of state, non-state, military, civilian, public and private actors. They comprise governments, international INGOs, military establishments, international financial institutions (IFIs), PMCs and PSCs, intergovernmental organisations (IGOs), and the business sector. These actors make up complex, mutating, and stratified networks, all interlinked, while at the same time representing different flows and groups of authority, expertise, and agendas. Together, these networks form what Duffield calls global liberal governance networks, the purpose of which is to ensure stability,[46] and thus give peace a chance.

Such complexes are strategic in that they are following a radical agenda of social transformation in the interests of global security.[47] These new complexes have expanded to include networks of strategic governance relations concerned not solely with development but also security, while at the same time they have sought to privatise and militarise their activities; a distinct break from the way previous complexes worked. Then, such complexes were solely concerned with managing development through aid agencies such as Oxfam, and IGOs such as US Agency for International Development (USAID) and the DFID.

New wars and strategic complexes of both types share certain common structural and organisational forms. The network of actors from both groups is increasingly private agents, rather than government representatives, working outside conventional areas of competence defined by governments. Private companies and charities, for example, undertake mine clearance tasks. Each group of actors is learning how to project power in new and imaginative ways. The use of private security companies by all the actors is a prime example of this. New wars and strategic complexes have blurred conventional distinctions between governments, armies and the people. In the case of new wars, the clear division of labour between each of these groups no longer resembles the Clausewitz trinity in the same way as in Western states. New wars and strategic complexes are also the product of market deregulation and outsourcing of nation state competence. Both groups reflect an emerging way of war that is very different to the earlier wars of the twentieth century.

The function of private security

Before explaining the purpose behind the use of private security in strategic complexes, we need to explain the difference between everyday private security and the private security offered in new wars. Everyday private security is the use of private commercial security to protect commercial assets, for example diamond mines and oil refineries, but also the assets of NGOs. Thus, not only does the security offered come from the commercial sector, the purpose behind using the security is commercial or, in the case of NGOs, reflects commercial concerns over the replacement of assets. The security offered in new wars is still commercial security, but it is military in nature. The purpose of this type of security is to support a ruler of a weak state engaged in fending off challenges to his authority by opposition groups, a consequence of which has been to undermine in certain cases any remaining legitimacy the ruler may have.

The role of PMCs in strategic complexes

The following four sections discuss the role of PMCs in strategic complexes. The first section offers a general overview of the position of PMCs in this arena. The remaining sections then discuss the specific roles PMCs have taken in security sector reform programmes, and with other actors, in particular multinational corporations and NGOs, in supporting businesses and development programmes.

All three sections provide an explanation of how PMCs have fitted into certain environments while dealing with certain situations in the development–security arena, and the tension this has created for other actors working in the same arena.

The wide range of services private security companies are able to provide has not been lost on political elites, multinational corporations, or NGOs. As mentioned earlier, in many cases, political elites have come to rely on such protection to be able to operate effectively, more so for rulers who have sought to commercialise politics to retain power. In this respect, protection has played a crucial role in denying warlords and strongmen access to strategic resource-laden enclaves, which in turn can link them into the international economy. Without this link, they are unable to manufacture profits from the sale of strategic resources that are necessary to continue to fight. Instead, such profit can go to rebuilding war torn economies, while also paying for the introduction of reform. In this environment, where politics is commercialised, that is, turned into a business venture, rulers have little option but to control local commercial networks, lest their position becomes unsustainable as political rivals challenge their authority.

Multinational oil and extraction companies that operate in the resource-laden enclaves also need protecting from the threat of attack. In this particular environment, the nature of the relationship between multinational companies and PMCs is one of mutual benefit; each profiting from the other. However, individual contracts can cause friction between these groups, especially if a multinational signs a contract with a local enforcement agency to pay for the creation of a special unit to protect their installations, then appoints a PMC to take on that task.[48] In general, however, the type of service most PMCs offer these companies is defensive in nature concerned with protecting assets from theft or attack, and not counter-insurgency. On the other hand, no close working relationship exists between PMCs and NGOs. This has to do with the way each group understands security. This does not mean that NGOs have not benefited from the actions of PMCs; they have, especially in Sierra Leone. But any direct involvement with PMCs is understood to threaten NGO neutrality and is the reason why NGOs prefer to have as little as possible to do with PMCs.[49] The following sections look at these relationships in detail.

PMCs and security sector reforms

The role of PMCs and PSCs in SSR can best be understood within the context of the new strategic complexes that now make up the development/security arena that has occurred within developing countries. The idea of including the security sector within development programmes appeared to come from the Scandinavian countries and the Netherlands. These countries viewed restructuring of the security sector in a less parochial and ghettoised manner than had previously been the case.[50] In this respect, reform of a police force in a developing country was not seen as an end in itself, but was part of a wider process that sought to reform the whole criminal justice system. Moreover, in terms of restructuring the military, reform was to be linked to national defence management that included the legislative and executive, while in each case, a more holistic approach to reform was adopted.

Evidence of this conceptual and normative shift could be identified in the different courses run for African countries by donor governments. US military training teams, in particular, began to emphasise the importance of stable civil–military relations to the management of a country's national defence force. There, programmes were widened to include civilian defence officials, parliamentarians and in some cases representatives of civil society.[51] Such programmes included areas as diverse as military training, defence budgeting, understanding the role of the military within a democracy, and logistics management to mention only a few areas. In the case of the UK, the MOD extended the remit of British Military Assistance Training Team (BMATT) in South Africa to assist the country in the creation and consolidation of its executive mechanisms of civil oversight over the armed forces, while it was the DFID that placed the issue of SSR on the international agenda in March 1999.[52] Its inclusion within the development arena was a result of donor thinking on the relevance of SSR on development issues and was achieved in stages.

The reasoning behind the need to include SSR in the field of development was based on the recognition that, according to Hendrickson, 'excessive security spending may also absorb scarce public resources that would be better used in other sectors contributing to poverty alleviation. Because security sector problems tend to be a symptom of broader social, political and economic challenges facing poorer societies, there is a strong argument for adopting a more holistic approach to development that incorporates security sector concerns'.[53] The following features characterise SSR, while also differentiating it from the security sector approach adopted by donor governments during the Cold War:[54]

- It has a clear normative and practical commitment to a development agenda;
- Its normative content is exemplified by its commitment to contextualise SSR within the ambit of the consolidation of democracy, the promotion of human rights, good governance and the creation of a culture of accountability and transparency in the management of security sector processes;
- The preparedness of SSR strategies to countenance a much higher degree of local ownership of the process than has hitherto been the case.

The introduction of SSR programmes was a big step forward in helping to resolve some of the major developmental problems facing developing countries. Such programmes, however, still face major challenges that will have to be overcome if the programmes are to succeed in their objectives. Nor are such challenges the prerogative of a single actor any more, but need the cooperation of all the actors working in the area of SSR.

The traditional approach to SSR was for Western governments to supply retired officers, government officials, or experts in areas such as water, agriculture, education, and trade and commerce, to advise on technical matters. The person was employed on a contract paid for by his or her own government. By the 1980s, private firms and consultancies were also starting to move into this area of work. Some firms even came from the security sector. As Ashington-Pickett explains, Control Risk Group started to supply security advice to development countries as long ago as 1982, while private security involvement in SSR is not at the same stage of development, but differs greatly between Western countries.[55] The US PMC MPRI, for example, has been undertaking military and reform programmes in the Balkans and other parts of the developing world for the last decade,[56] while in the UK foreign military training remains the responsibility of the British Army. MPRI has also undertaken reform within the area of justice and law enforcement. In the UK, this responsibility has been outsourced to the private sector, but not necessarily to PMCs and PSCs. Instead, the public sector consultancy business has taken on this area of work.

Companies such as Atos Consulting, whose expertise is in generic public sector reform, have for some time been involved in reforming security and justice systems in developing countries. One of the company's biggest projects in this particular area was undertaken in Jamaica, where it was asked to review three large ministries, one of which was the Ministry of National Security and Justice under which the Jamaican National Defence Force sits.[57] The company's expertise in public sector reform has enabled it to undertake work in wider public security sector reform projects, while in the last few years the post-conflict agenda has been added to it, since conflict often has its origins in the failure of the justice sector.[58]

Much of the work undertaken by public sector consulting firms is at the strategic level and is designed to promote long-term stability, through helping to building local capacity and competence. In this respect, there is a substantial difference between immediate post-conflict tasks that PMCs and PSCs are now undertaking in places such as Iraq and other conflict zones and long-term reform programmes now being implemented in places such as the Balkans by public sector consulting firms. Neither is this distinction lost on the companies. According to Ashington-Pickett,

> While security companies have an array of skills to offer SSR programmes, before any long-term reforms can be implemented local forces need to be disarmed, demobilised, and then reintegrated (DDR) back into society. This process should occur soon after the fighting has stopped and in the order just mentioned. It is also an area of work that PSCs can assist the international community in. After all, security companies have a range of skills at their disposal that can be deployed immediately to help with stabilising a country that has collapsed.[59]

Howard makes the same point;

> Most PMCs and PSCs are in the game of rapid response, plugging the gap where a country has fallen apart, while our job is to reform failing public

institutions including those concerned with justice and security. [Continuing, Howard explains that] it is still possible for both types of organisations to cooperate in long term SSR programmes. The nature of that cooperation could be a partnership, or a payment for the supply of certain services. For example, specific tasks such as training a developing country's police force or a local defence force may be given to a PSC with specialist skills in this area because their background is seen as more credible to the client than our background. The same is also true of military programmes. This leaves the public sector consulting firm to concentrate on wider SSR issues in the country.[60]

In the end, cooperation may be the best way to tackle the numerous challenges facing the international community as it attempts to reform the security sector of developing countries, particularly in Africa.

African militaries desperately need restructuring so they can deal more effectively with the new security threats they face. Many of these militaries need to be democratised. In this respect, they need to understand that they are responsible to the state, not the individual who holds the office of state. Furthermore, they must stay out of the political arena. To achieve both tasks, African militaries will need to be retrained to professional standards. The likelihood of this at present is very low, mainly because of the lack of money spent on them, and the poor training and wages this creates, which leads to poor morale, and because of broader social–political issues. At the same time, these problems are not insurmountable if the responsibility for reform is shared among the different international actors.

There may even be a role for African PMCs and PSCs in SSR programmes working alongside international agencies. It is more likely, however, that they would form a partnership with either European or US PMC or PSC. In doing so, the latter group may find it easier to sell military training courses run by former African military personnel to African militaries, as opposed to those run by European or US military personnel. After all, EO was also involved in retraining African militaries, undertaking contracts to train part of the Angolan military as well as part of Sierra Leone's military force. In doing so, it was hoped these countries' military forces would act more effectively to end their country's civil war and bring about a return to stability. Unfortunately, these efforts were never recognised by the international community, notably because the reforms took place during the civil wars instead of after the wars had concluded.

In many ways, South African PMCs and PSCs may be better placed to conduct military reforms in Africa than companies based outside of Africa. Their local knowledge and experience means they may have a greater understanding of the type of problems African militaries face such as lack of equipment, poor morale, and poor training. Resolving these issues may require an awareness and sensitivity to local culture that a local PMC may have, as EO demonstrated in Angola, that may be lacking in Western companies. In Angola, for example, EO used its local knowledge to help local people overcome the problem of poor sanitation by purifying water for them and then not charging for this service. As Barlow explains,

We had to purify water for ourselves anyway, straight out of the Longa River. The spin off from that was that the recruits saw that we were actually being good to the locals. The locals saw us, in a sense, as saviours, because this was the first time anyone had ever done anything for them free, and we did not expect anything in return for it. Although we knew that if we had built up a good rapport with the locals, they would warn us of any hostile activities in the area because we would become their friends, and they would not want to lose us, and they were getting something free from us. [61]

Consequently, a good rapport was established between the recruits and EO personnel making their job easier and thus more effective.

The company was also in a position to help in retraining the Sierra Leone military. And if SSR had been introduced immediately after EO's successes, the people of Sierra Leone may not have had to face three more years of fighting. Unfortunately, tension between the company and sections of the international community prevented any chance of this happening. Instead, the company faced a barrage of criticisms from human rights groups, while the UN refused to have any dealings with them, creating, as Shearer has argued, further obstacles to the successful implementation of the peace agreement. [62]

In the end, local PMCs can, with the right support, manage the more specific SSR programmes and thus are likely to want to challenge PMCs from Europe and the US for those contracts. Unfortunately for them, they will face a number of problems. European and US PMCs are likely to poach their employees for their local knowledge and experience by offering a better remuneration. Moreover, they can afford to offer greater incentives, as well as applying political pressure on local governments to ensure they receive the contracts to undertake SSR programmes. Certainly, with the type and amount of resources available to them, local PMCs will find it hard to compete. Even so, no one should underestimate their ability to carry out this type of work. After all, they do have a wealth of local knowledge and experience at their disposal.

PMCs and their relationship to multinational corporations and international organisations

The commercialisation of politics by rulers, taking on warlord tactics to stay in power has complicated the use of PMCs in the development/security arena. [63] For those PMCs offering commercial security, this has led to claims that all they are interested in is securing resource-laden enclaves to boost their profits rather than protecting rulers. [64] In reality, however, the situation is more complex. The ruler wants these areas secured to deny the profits to opposing strongmen, while selling the resource to buy weapons and loyalty which in turn allows him to continue to rule. Thus, for PMCs, the priority is to establish a secure environment by enhancing institutional capabilities that allow rulers to act. This leads to the mineral extraction companies benefiting because of the action of the PMC. Furthermore, problems can arise if the relationship between a military company and a company contracted to exploit the resource is too close,[65] as in the case of individuals simultaneously holding positions in both PMCs and resource extraction companies.[66] In this situation, the security interests of the ruler may be marginalised by the PMC in favour of the security interests of the extraction company. In the end, some type of structural adjustment between military and extraction companies will be necessary to ensure the interests of the ruler always come first.

PSCs on the other hand may be better placed than PMCs to take on commercial contracts including security. PSCs see themselves as wholly legitimate having established good relations with numerous juridical states. This is reflected in the security activities they have undertaken for Western governments, as well as international organisations. ArmorGroup for example, includes as its clients UN agencies, the UK government, the EU Commission, the USAID, and many more government and international organisations.[67] The roles they perform include providing risk analysis and planning, security for facilities and operations on the ground, global responsiveness, protection of remote sites and lines of communication, and providing guards.[68] In developing countries, they already provide security solutions and protection to embassies, multinational corporations, and occasionally NGOs.

Whether PMCs[69] or PSCs are used, private security is now an essential component for foreign companies and international organisations working in developing countries that have recently emerged from civil conflict. Their presence is necessary if outside investors are to be confident that their money is safe. Without this guarantee, they are unlikely to want to invest, leaving this responsibility to national governments, or institutions such as the World Bank. Whether developing countries can function more easily with commercial investment that links them to the international economy through institutions and not criminal networks is questionable. What is not in question is the need for developing countries to manage their security properly to have any hope of attracting outside investment. It is here private security is playing a crucial role, providing security expertise to local governments, while investor confidence increases as a consequence of improving security arrangements.

PMCs and NGOs

It is in the area of development assistance that a potential role for PMCs exists. In relation to development networks, we are already witnessing increased interaction between military and security actors with international organisations and NGOs. There even appears to be a thickening of development networks that now link UN organisations, military establishments, government departments, NGOs, and PMCs. After all, only the military or PMCs are equipped with the means to ensure the protection of those who deliver aid to complex humanitarian disasters and the recipients of that aid. Moreover, normally viewed as neutral by warring sides, NGOs in particular are now more vulnerable to violent attacks. Attacks against aid workers have grown at an alarming rate in the last few years.[70] Thus, there appears to be a potential role for PMCs to assist NGOs in their work in filling the security gap left by host states' inability to provide security for NGOs. The complex nature of development work, the fact that much of it is carried out in or around war zones, or in countries where war has been a semi-permanent feature of life, places PMCs in a particularly strong position for undertaking development assistance because of the skills they are able to offer.[71]

Even though the international community rarely acknowledged their efforts, EO did demonstrate, using the array of skills they held, how useful they could be to local communities struggling to survive. As mentioned earlier in the chapter, while operating in Angola, the company purified water for the local villages. In Sierra Leone, they undertook to fly children orphaned by the RUF back to Freetown, where a Catholic funded organisation was waiting to take care of them. Had the children been left in the interior of the country, they would have undoubtedly gravitated to the RUF for food, turned to prostitution, or died.[72] Other EO employees helped locals grow their own food, though EO employees also benefited.[73] While working in Angola, the company undertook to fly medical supplies into the country. Even though their shelf life had expired, the supplies would have undoubtedly improved the quality of life for some, if only for a short period. The supplies came from hospitals in South Africa. As Barlow states,

> We started with lots of projects to get money to help the Angolans. We released a compact disk record, which we had artists record and was later sold to raise money. By the time we left Angola there were approximately nine clinics running, which were supported by medicines we got from hospitals in South Africa – time expired medicines – that before they were destroyed we got them and flew the medicines to Angola. We did not sell it to the Angolans, we gave it to them. It had many other positive effects for us. We took many South African companies into Angola, helped them establish business there in order to boost trade between Angola and South Africa.[74]

Such acts of generosity are undertaken for a reason, a fact not disputed by the company. In the case of EO, they improved the position of the company among local communities, and that in turn would result in them receiving more intelligence information from the local population.

Even though these acts made very little, if any, difference to the situation in each country, they highlight an important point, that EO personnel were very much aware of the developmental problems the majority of people living in war torn countries face. Furthermore, since most, if not all, of EO's employees were African, and may even have faced very similar difficulties, they were able to empathise with local populations.

More importantly, NGOs could benefit from the mass of experience contained within the management of PMCs in dealing with the type of complex security issues that NGOs regularly face in conflict zones. At the same time, if a PMC does not have the expertise, they can use their database to identify who does have that experience. The database is also likely to hold the details of individuals with specialist knowledge of particular conflict zones. With all this information available, the PMC is able to approach complex security issues in a holistic way, not as discrete issues. Passing that experience on to NGOs could improve their level of productivity, while enhancing the security arrangements of their staff.

There are also advantages for both groups in using a contractual arrangement to resolving complex security issues. No two war zones are the same, thus requiring NGOs to take a flexible approach to any security issue they face. As the social and political environment of a war zone changes so do the security threats. Being able to respond to such changes is important for NGOs. A situation may occur for example, that results in the rapid deterioration of the security environment that requires the evacuation of all NGO staff. Such a situation could be included in a contract, and that requires a PMC to draw up a contingency plan for the evacuation of the staff. At the same time, the NGO would not pay for any of this, apart from the cost of drawing up the initial plan, unless the plan was put into action. Under this type of arrangement, an NGO would save paying for excessive security measures, while knowing a contingency plan was in place if the security situation deteriorated to a level where their staff faced an unacceptable security threat.

Even so, there are also problems associated with using PMCs to support humanitarian operations. The use of force in these operations can easily undermine the humanitarian principles that guide them and thus bring into question the legitimacy of such action. As International Alert explains,

> if personnel are armed or are from a military background, then aid agencies might be perceived as being part of the conflict rather than mere bystanders there to assist its victims. This challenges the notion of impartiality and neutrality of humanitarian action and its identity as a civil movement.[75]

Even the slightest perception of any association with a PMC can do irreparable damage to their public image.

The use of PMCs may also create a perception among donor agencies and the victims of humanitarian disasters that the protection of humanitarian staff is the main goal of aid agencies and not the safety and wellbeing of the victims. This may perversely affect the relationship between aid worker and victim notably by distancing the two groups. Security should not be understood as a goal in itself, but as a means to an end, the objective being to help reduce human suffering through the immediate and impartial delivery of humanitarian aid. It is also possible that by choosing the 'hard' security option, humanitarian and political interests may become blurred, further increasing the risk to humanitarian staff from different warring factions. More important for the victims of humanitarian disasters is the impact of using armed guards to deliver humanitarian aid on the security of local communities. The use of PMCs to protect aid deliveries may be seen as an attempt to win the confidence of the locals in the hope of gaining important intelligence from them.

Although there have been exceptions, in the past it has been the policy of aid agencies not to use PMCs. Aid agencies have occasionally relied on the services of PMCs when they have felt they lack a sufficiently comprehensive picture of the security problem in order to make a judgement about the most appropriate response.[76] But there is still some way to go before PMCs become an integrated part of the aid agency network. Convincing NGOs that they can gain substantially from the services PMCs can offer will take time, and even then PMCs can not be sure of success. Many NGOs are still suspicious of PMCs, and have rigorously contested the inclusion of PMCs in the developmental network because they may compromise their neutrality. This has to do with the way local communities and warring factions understand traditional security, which they base on acceptance, and that depends on a perception of impartiality, while employing a PMC may compromise that perception.[77] Merging of development and security has thus made it increasingly difficult for NGOs to operate. Suggesting they engage with PMCs at this stage would be asking too much, especially when differentiating PMCs from a mercenary outfit is still problematic. Licensing PMCs may be a way of achieving this separation.

Conclusion

As a result of the shift from patronage politics to warlordism, many countries in Africa have slipped into civil war as political elites have fought over control of resources. Unable to prevent these wars, Western governments have either ignored them or sought alternative structures of conflict resolution, linking and integrating state and non-state actors, public and private organisations, and military and civilian organisations to bring these wars to an end. As such, strategic complexes are now important formations in confronting new wars and the merging of development and security, facilitated by the privatisation and subcontracting of former state development and security responsibilities. Thus, innovative networks connecting government, NGOs, and the business sector have not only begun to emerge and consolidate, but are now assuming responsibility for securing peace.

The privatisation of violence by members of strategic complexes is the most controversial aspect of their response to new wars and the convergence of development and security. With occasional exceptions as previously noted, the West's reluctance to deploy military force to end new wars has meant NGOs, international organisations, and the business sector turning to alternative measures to protect their organisations and create a secure environment in which to work. This has seen them turn to the private sector for help. The employment of private security companies to guard facilities, including oil refineries, diamond mines, humanitarian convoys and food aid, is now a common occurrence for members of strategic complexes. Without the security offered by PSCs, many members are unlikely to fulfil their intended objectives. As important is their experience in the wider military and security arena that potentially makes them suitable to undertake SSR in developing countries on behalf of donor governments. More controversial is the provision of direct military services for rulers, even outside the realms of donor control, in conflict resolution. In Angola and Sierra Leone for example, EO managed to tip the balance in favour of the government and so resolve the conflict.

Others have argued that such gains have usually been temporary using legally questionable methods. Furthermore, using PMCs is tantamount to an unregulated privatisation of war.[78] Reno, on the other hand, points to the fact that rulers within weak states face real internal security threats from rival strongmen. They will meet these threats using whatever means they have at their disposal including using income from natural resources to hire PMCs.[79]

The role of PMCs in the new strategic complexes is far from straightforward, particularly so in the area of SSR. While the US has forged ahead using companies such as MPRI to undertake reform programmes for armies in the developing world, European countries have been hesitant to follow this approach. There are signs that this may change or indeed already be changing as European militaries become overstretched as a result of increasing commitments. It is certainly the case that many security companies in the UK and in other parts of the EU have a wealth of experience at their fingertips and would be willing to place such experience at the disposal of a government or an international organisation responsible for funding SSR programmes. But the political will needs to be there for this to occur.

Such a move could even draw in security companies from countries such as South Africa as partners to European and US security companies. EO demonstrated in Angola and Sierra Leone that a South African PMC has the knowledge and experience to participate in SSRs. Furthermore, the company's African roots may have made them more sensitive to local concerns than maybe a European or US Company would have been. The company certainly demonstrated an ability to train local military forces in Angola and Sierra Leone. EO also had a very clear understanding of the types of problems many local communities face, and attempted to support them through local initiatives however limited these initiatives were. The company was able to bring the fighting in Angola and Sierra Leone to an end, creating the necessary stable environment for security sector reforms to take place, even if they were never able to resolve the issue of legitimacy.

MPRI, on the other hand, faces no legitimacy problem, especially in America. The company is thus able to undertake work for the US State Department, acting as an extension of US foreign policy. The company, however, does not undertake the type of combat support operations EO undertook in Angola and Sierra Leone. Instead, the type of work it undertakes is to train, or in certain cases retrain, the militaries in developing countries. MPRI provides training for African militaries through the ACOTA programme. ACOTA is a US State Department coordinated interagency programme that works with African states and other allies to develop and enhance peace operations and humanitarian assistance capabilities among selected African armies.[80] The impact of this is that in the future the US military will be able to undertake combined operations with African militaries more easily, thus reducing the political risk and cost to the US government. There is also the possibility that once trained, African militaries will undertake the combat aspect of any operation in Africa, while the US military provides logistical support. This can only happen though once African militaries have been trained to work alongside the US military. Considering the scale of America's war on terror, this last point is particularly relevant since the US has neither the manpower nor resources to engage in every conflict that terrorists may be involved in. For the US government, sharing the burden in this way is only sensible.

The relationship between PMCs and multinational corporations or international organisations has until recently received less attention than that between PMCs and governments. At the same time, the relationship has become increasingly active at the international level. PMCs are taking on increasingly important and innovative roles for a range of clients in war zones. Such roles include guarding the properties of multinational corporations and protecting personnel. This trend is set to expand for a variety of reasons. The changing competence of nation states, market deregulation, the growth of privatisation, and the downsizing of military establishments have all contributed. Finally, the most contentious relationship is between PMCs and NGOs. NGOs have historically distanced themselves from the security services offered by PMCs, refusing to engage with them apart from in exceptional circumstances. But at the same time, they have discretely used private security to advise them on security issues at the strategic and operational level, while local security is used to physically protect their supplies because it carries less risk of their aid being politicised. They have also benefited from the stable environment that the actions of PMCs have created.

The privatisation of security is suited to new wars. While still evolving and facing an uncertain future, PMCs are likely to expand their role, especially since the current trend points to them being legitimated through the introduction of regulations. PMCs can help to address tensions and needs Western states are finding difficult to encompass. It is unlikely, however, that the type of operations conducted by EO in Angola and Sierra Leone will be repeated, unless sanctioned by a major power or the UN. At the same time, the commercial sector also needs protection. PMCs provide one solution to these differing requirements. Whether they can provide a lasting solution is a different matter.

7

REGULATION AND CONTROL OF PRIVATE MILITARY COMPANIES

The legislative dimension

Introduction

As the previous chapters have explained, the demand for private military and security services is likely to increase in the future and therefore PMCs will continue to have an impact on international security. But while the number of PMCs operating globally continues to grow, the legal arguments surrounding their role on the international stage have been marginalised, if not ignored. As a result of this, their legal position remains ambiguous. As noted earlier, PMCs are private companies selling military and security services. In the past, international law termed those selling such services mercenaries and prohibited the activity. Unfortunately, the international Conventions to regulate mercenary activity have proved ineffective, especially when applied to PMCs. The very narrow definition of mercenary found in the 1977 Additional Protocols I and II to Article 47 of the Geneva Convention (1949)[1] and the International Convention against the Recruitment, Use, Financing and Training of Mercenaries (1989)[2] have made it easy for mercenaries, let alone PMCs, to avoid meeting the full criteria of the definition and thus prosecution. These Conventions reflect international tensions between the West and parts of the Third World over what they see as the West's willingness to tolerate mercenary activity beyond their borders. Furthermore, the case of British mercenary Simon Mann appears to support the argument that the Foreign Office knew about his attempted plot to overthrow the government of Equatorial Guinea before it happened, but failed to intervene to stop it.[3] These international Conventions are therefore of little use in regulating the actions of PMCs by establishing their legal position in international law.

The inability of the international community to establish the legal status of PMCs effectively places the issue at the national level. Furthermore, this approach does not require the cooperation of other countries and is therefore much easier to implement. It does not depend on international negotiation, which frequently means that the introduction of any law represents nothing more than the lowest common denominator and thus is ineffective. National laws, on the other hand, do not suffer from this problem and are therefore more effective, as long as the government has the will and resources to apply them. Once again, the fact that the mercenaries accused of plotting to overthrow the government of Equatorial Guinea were charged under domestic law, while no mention was made of the International Convention for the Recruitment, Use, Financing and Training of Mercenaries, supports this point. Even so, as the chapter explains, some national regulations designed to curb the sale of military and security services are also deemed too weak to be effective, while other national regulations entail significant challenges for any government, especially since the activity normally takes place overseas.

The chapter first briefly details the reasons why governments need to regulate the sale of military and security services offered by PMCs and PSCs. Next the chapter focuses on the Foreign Enlistments Act 1870 and the US Neutralities Act of 1939 explaining why these national regulations have been ineffective in controlling such sales before focusing on the US ITAR and South Africa's Foreign Military Assistance Act 1998. Each of these two approaches to controlling the sale of military and security services is discussed including their historical development and the advantages and disadvantages of each system. Finally, the UK government's Green Paper that sets out some of the options available to those governments that want to regulate the activities of PMCs and PSCs is examined in detail.

Reasons for regulating PMCs

There are eight important reasons why PMCs should be regulated:[4]

- To ensure they do not adversely impact on peace, security and conflict resolution.
- To ensure their use is both legal and legitimate and does not contravene human rights;
- To ensure they do not undermine government policy;
- To prevent them causing economic damage to their commercial clients;
- To ensure they are made accountable both for their actions and for those of their employees;
- To make certain they are as transparent as possible;
- To prevent them shifting between legal and illegal pursuits;
- To guarantee they do not in any way undermine the sovereignty of states.

Just as important, however, is what a government would hope to achieve by regulating PMCs. Only after answering this question will a government be in a position to choose the regulatory option that will allow PMCs to function efficiently, while at the same time protecting human rights. Indeed, if a government's reasons for regulating the industry do not match the aims of the companies operating within its own borders, or simply fail to acknowledge those aims, then PMCs will face a very difficult future which may even force some companies to operate from overseas in order to survive. In this respect, the purpose of regulating PMCs should not compromise the business practices of companies as long as those practices do not compromise international standards on human rights. Ensuring this balance is maintained will be no simple task for those responsible for regulating PMCs.

Problematising national regulations

At the national level, the majority of domestic legislation is simply out of date. Such legislation does not accou[...] nply deficient in legal terms to define and reg[...] most obvious in the UK but also in the US. T[...] n Enlistments Act 1870[5] is preoccupied wit[...] es during the previous decades and the mea[...] vailable. This is not surprising considering tl[...] ıs designed to regulate and control a social e[...] ppeared more than 100 years ago. The Act, [...] 3ritish subject without licence from Her Majesty to enlist in the armed forces of a foreign power at war with another foreign power which is at peace with the UK, or for any person in Her Majesty's Dominions to recruit any person for such forces, has been almost impossible to enforce.[6] Moreover, none of the prosecutions under the Act, or its predecessor of 1819, have been concerned with enlistment or recruitment. Nor were those who enlisted to fight in the Spanish Civil War ever prosecuted. The Director of Public Prosecution at the time abandoned the idea because of the problems associated with assembling evidence of an activity taking place abroad.[7] The Diplock Committee, which was established to investigate the actions of British mercenaries who fought in the Civil War in Angola during the mid-1970s, examined the question in some detail and concluded the Act was ineffective and should be repealed or replaced, but no action was taken.[8] The same situation exists in the US with the Neutrality Act of 1939. The law is only concerned with prohibiting the recruitment of mercenaries within the US and does not concern itself with the sale of military services.[9] This is not surprising given that the Act was again a reaction to a set o[...] at that time have to take into account the sal[...] use they did not exist. Today those circumsta[...] ferent. Even attempts by the US government [...] een wholly successful. As Singer points out, [...] Jurisdiction Act was intended to fill the gap, b[...] ng in the US military outside of the United St[...] contractors working directly for the US DOD [...] cover those contractors working for another US agency, or a contractor working abroad for a foreign government or organisation. Therefore, if an American contractor working for a PMC commits an offence abroad, unless they are working for the US DOD on US military facilities, they may escape prosecution, the only alternative being for the host nation to prosecute.

International traffic in arms regulation (ITAR)

While it may be difficult, or simply not possible for a government to use national legislation to prosecute its citizens working for a PMC abroad selling military services, the US government is able to regulate certain of these services using other legislative means. The purpose of ITAR is to control, *via* regulation, the export and import of US defence articles and defence services. The regulations are primarily administered by the Director of the Office of Defense Trade Controls, Bureau of Politico–Military Affairs, Department of State.[11] Generally, only US persons and foreign governmental entities in the US may be granted licenses and only if the applicant has registered with the Office of Defense Trade Controls. At the same time, the Arms Export Control Act (22 U.S.C. 2778 (a) and 2794 (7)) provides for the designation of those article and services deemed to be defence articles and defence services and thus designated to constitute the US Munitions List. Such designations are actually made by the Department of State with the concurrence of the DOD.

A US PMC would need to be registered with the Office of Defense Trade Controls before it could apply for a licence to export defence articles or defence services to a foreign military force, National Guard, or police force. Such article and defence services include certain kinds of technical data, military equipment, and defence services. In respect to US PMCs it is the export of defence services that would most likely affect them the most. This is because they are defence service providers and not defence equipment manufacturers. They therefore have to apply for a licence before they can furnish assistance to foreign governments including helping to repair, maintain, and operate defence articles. A licence is also needed before they can undertake military training of foreign units and forces, regular and irregular, including formal and informal instruction. Neither can they undertake to give military advice without a licence. At the same time, ITAR gives official approval to such transfers. This is not a complete picture of the licensing requirements placed on US PMCs, however it shows to some degree the level and type of control the US government has over their activities.

As mentioned above, these regulations are designed to prohibit the export and sales of defence articles and defence services to certain countries, notably those prohibited by UN Security Council arms embargoes and countries which the Secretary of State has determined to have repeatedly provided support for acts of international terrorism and thus run contrary to the foreign policy of the US. Such countries include Iran, North Korea, Sudan, Syria, Libya, and Iraq up until the intervention of US and UK troops.[12] How successful ITAR has been in ensuring the activities of US PMCs do not compromise these aims is hard to say. As Professor Deborah Avant explains, '[t]he Defense and State Department offices that have input into the process vary from contract to contract, and neither the companies nor independent observers are exactly clear about how the process works'.[13] Singer also believes ITAR offers only minimal regulation while providing official approval, which might supersede international regulation.[14] Furthermore, under present US law, a US PMC working abroad does not have to notify Congress if the contract does not accede US$50 million, which many contracts fall under or are simply broken up to ensure this. However, the company will still require a licence if they are exporting defence services covered by the US Munitions List. Finally, once a licence has been issued to a US PMC, there are specific oversight requirements to ensure the company is adhering to the terms of the contract. Generally it is the responsibility of the most senior US embassy official in the contracting country to ensure a PMC or their activities are monitored. But, while such a person may have a general responsibility for oversight, no one is actually dedicated solely to this responsibility, while according to Singer, many embassy officials see this as contrary to their job.[15]

Regulation of Foreign Military Assistance Act[16]

Only the South African government has so far introduced a law to prohibit their citizens from rendering military training to nationals of a country in conflict. The government decided in 1995 to pass an Act which would control unauthorised foreign military assistance to any State in conflict. The decision was necessitated due to the negative international coverage regarding the activities of EO in Angola and Sierra Leone, and other South African firms operating in Africa and beyond. For former apartheid soldiers, selling their military skill to foreign governments earned them considerably more money than working at home. The Regulation of Foreign Military Assistance Act was passed in September 1998, making the country one of very few in the world to outlaw mercenarism and regulate foreign military assistance.[17] The Act aims to prevent South African companies and citizens from participating in armed conflict, nationally or internationally, except as provided for in the Constitution or national legislation. The Act prohibits any mercenary activity, defined as direct participation by a combatant in armed conflict for personal gain, while also prohibiting the rendering of military assistance, or any offer to render such assistance unless authorisation has been obtained from the Minister of Defence who, in consultation with the National Conventional Arms Control Committee, can either refuse or grant an application for authorisation subject to conditions they can determine and can at any time withdraw or amend any authorisation they have granted.

The new law, however, is problematic. As with international law there is the problem of definition. The Act defines foreign military assistance in very broad terms and includes the following:[18]

- Advice and training;
- Personnel, financial, logistical, intelligence, or operational support;
- Personnel recruitment;
- Medical or paramedical services;
- Procurement or equipment;
- Security services for the protection of individuals involved in armed conflict or their property;
- Any action aimed at overthrowing a government or undermining the constitutional order, sovereignty or territorial integrity of a state;
- Any other action that had the result of furthering the military interest of a party to the armed conflict, but not humanitarian or civilian activities at reliving the plight of civilian activities in an area of armed conflict.

By extending the legal net as wide as possible, the government has sought to control every form of foreign military and security services. Unfortunately, the definition can include the activities of actors such as aid agencies involved in conflict prevention, and humanitarian aid. This point has already been acknowledged by some of the organisations that have sponsored the bill. Their primary concern 'is the bill's potential for collateral regulation of *bona fide* activities which have nothing to do with mercenary activities, but may fall under the very wide umbrella of foreign military assistance', and as a result of which some South African aid agencies may find some of their activities, such as humanitarian demining, subject to licensing by the government.[19] This could lead to delays in delivering humanitarian services, or prosecution if the service is undertaken without the appropriate licence in place.

Another problem for the South African government is that by requiring PMCs to obtain a licence for each contract, the government may be seen as officially sanctioning such contracts. By making it a legal requirement for PMCs to obtain licences for contracts, the South African government could now be legally accountable for the actions of South African PMCs by a foreign government. However, this situation has not yet been tested in a court of law. Whether this would allow PMCs to escape those international legal controls discussed earlier is not clear. In the end, it may make little difference since such controls have normally proved ineffective in curbing the activities of PMCs.

Whatever the problems are with the Foreign Military Assistance Act, the intended impact it was expected to have on the private military/security industry has so far been marginal. The Act has not prevented former South African soldiers or police officers from working in Iraq for example. It is estimated that 1,000 South Africans work as security personnel in the country at present.[20] Neither has it stopped them from operating in other parts of Africa.[21] Moreover, South Africa's most controversial PMC, EO, moved their headquarters overseas to escape the Act before the company ceased trading altogether, thus highlighting one of the problems associated with national approaches to regulation, in that it requires governments to coordinate their efforts with other governments.

The only other government to have seriously considered regulating PMCs is the British government. At present it is examining different licensing possibilities and has set out a number of options in a Green Paper, which is discussed in detail below. As with ITAR and the Foreign Military Assistance Act, the Green Paper broadly accepts the need for a licensing regime to control the activities of UK PMCs. So far, however, the Green Paper, which originated in a request from the Foreign Affairs Committee of the House of Commons and published in 2002, is the only response the British government has made towards regulating PMCs. It has yet to make any substantive move in this area and may well delay such a move until the political climate is more receptive to the idea of using PMCs for military/ security related tasks instead of state troops.

The Green Paper

In the foreword to the Green Paper, the Foreign Secretary, Jack Straw, made a number of observations. First, the control of violence is one of the fundamental issues in politics, perhaps the fundamental issue. Second, the post-Cold War world has given rise both to new problems and new opportunities. Third, the demand for private military services is likely to increase and, fourth, today's world can be considered a far cry from the 1960s when private military activity usually meant mercenaries of a rather unsavoury kind being involved in post-colonial or neo-colonial conflicts.[22] The Foreign Secretary then made two very important suggestions. First, that PMCs could constitute a viable alternative to national forces when the international community lacks the political will to respond to crises and other humanitarian contingencies. This view is very different from that held during the Cold War. Then, the blocking of Soviet intervention was the driving force behind Western intervention. With the end of the Cold War, this situation no longer exists. Instead, public opinion, financial costs, and the impact of intervention on an overstretched military, determine whether the UK responds to such crises. Second, that the government consider the option of a licensing regime to distinguish between reputable and disreputable PMCs. Such a regime would support and encourage reputable PMCs while, as far as possible, eliminating the latter.[23] But before the government takes any decision as to whether PMCs should be allowed to operate, they will need to consider very carefully a number of key issues. These issues include public–private sector interests, accountability, transparency, and regulatory enforcement.

Below are outlined the six regulatory options discussed in the Green Paper. While the government has not expressed a preference for any of the options, including the option of doing nothing at all, the language used in the Green Paper suggests that some kind of regulatory regime is being considered. Jack Straw, the Foreign Secretary, was reported in *The Financial Times* as wanting to see introduced a case-by-case licensing system to regulate PMCs in an attempt to avoid a repeat of the 'arms to Africa' affair.[24] The same feeling exists among members of the industry, analysts, activists, and academics who all agree about the need for some type of regulatory framework. The challenge facing the UK government is to establish one that is enforceable and that will satisfy key demands: defending the public interest, the interests of the recipient government and its people, and allowing companies to continue with their legitimate business activities. If these demands are to be met, the government will, according to Menzies Campbell, the Liberal Democrat spokesperson on foreign affairs and defence, have to strike a balance between liberty and responsibility, only intervening when there is a clear need to monitor or protect the public interest.[25]

140

Six options for regulation

Six options for regulation are outlined in the Green Paper:

- A ban on military activity abroad;
- A ban on recruitment for military activity abroad;
- A licensing regime for military services;
- Regulation and notification;
- A general licence for PMCs and PSCs;
- Self-regulation: the introduction of a voluntary code of conduct.

A ban on military activity abroad

According to the Green Paper, 'an outright ban on military activity abroad could only be achieved by either amending the 1870 Foreign Enlistment Act or by independent legislation'.[26] More importantly, such a ban would have an immediate impact on an activity that many citizens find unacceptable, while Parliament would decide whether the legislation would cover all military activity or a limited range, for example combat operations.

Enforcing an outright ban would be very difficult for a number of reasons. An outright ban on military activities abroad would raise the problem of definition. Trying to ban something as broad as military activity is very difficult. The government would have to explain what actually constitutes military activity. If the ban applied only to active participation in combat, would this include giving strategic advice or other vital support to military operations but one which did not actually include firing at the enemy? Then again, would such a ban include the provision of medical services to warring parties, compromising the activities of humanitarian organisations such as Médecins sans Frontières? Just as important, what would be the impact of such a ban on charities undertaking humanitarian mine-clearance work, which is certainly a military activity but one adapted for humanitarian purpose? One option may be to introduce a framework that permitted the operations of charitable organisations providing not-for-profit humanitarian aid or PMCs providing security to private firms on the one hand, but which banned the provision of direct support for offensive operations. In the end, such an approach would almost inevitably be open to allegations of inconsistency leading no doubt to grey areas being tested in the courts which could lead to clarification through case law or additional legislation.

Neither would it be easy to attain the necessary evidence for a successful prosecution in British courts since the activity takes place abroad. However, this should not be seen as an excuse not to introduce legislation to control the activities of PMCs working overseas, but reflects one of the main hurdles the legislation will have to overcome. Here the introduction of extra-territorial clauses into the arms export policies of the UK may be helpful.[27] Prohibition would also be an unwarranted interference with individual liberty according to the Diplock Committee.[28] In his report Diplock argued,

Any claim by the state to exercise control over the enlistment of its citizens as mercenaries for services abroad involves a restriction upon the freedom of the individual which, in our view, requires to be justified on grounds of public interest.[29]

In respect to changes in attitudes to international norms of behaviour that have taken place over the last 30 years, such a prohibition may be outdated leading to an opportunity to re-examine this particular claim. It may now be possible therefore to withhold an individual's liberty if the state felt that by working for a PMC the individual posed a threat to human rights.

At the same time, weak but legitimate governments would be deprived of much needed support of the type that the international community is either unable or unwilling to give. Given these problems, a more suitable response, therefore, would be to ban the recruitment of British citizens for military service abroad. However, such a ban is also problematic as the following section explains.

A ban on recruitment for military activity abroad

The Diplock Report, published in 1976 following British mercenary involvement in the Angolan Civil War, recommended that legislation concentrate on prohibiting activities that encouraged or facilitated the recruitment of individuals for military service abroad in specified armed forces. It did not recommend that such service itself should be made illegal.[30] Instead, the recommended legislation was aimed at preventing the recruitment of mercenaries. Persons undertaking such activity would then be liable for prosecution. The legislation would be in the form of an Enabling Act empowering the government when necessary to specify particular armed forces for which recruitment would be illegal.[31]

The advantage this approach would have over an outright ban is that it avoids those difficulties associated with legislating for activities that take place abroad, while enabling the government to intervene only when British interests are threatened. The proposal was primarily aimed at the recruitment of freelance mercenaries employed on a short-term basis. Unfortunately, freelance mercenaries rarely take notice of national law.[32] Furthermore, it is doubtful whether it would stop permanent employees of PMCs from acting as security consultants. Banning recruitment may even have an adverse impact on the industry, since the majority of PMCs tend to operate on a subcontractual basis.[33] Using subcontractors keeps running costs to a minimum by the maintenance of a small permanent staff and relying on networks of ex-servicemen to fulfil contractual duties. Trying to draw a clear distinction between subcontractors and freelance mercenaries is almost impossible and could deny ex-servicemen legitimate earnings.

Finally, neither might it prevent a PMC from transferring men legally recruited for one conflict moving to another.[34] On the other hand, the government may be able to prevent a potentially catastrophic intervention by a PMC. But it may not serve to improve the respectability of the industry. At the same time, PMCs could escape such measures by moving offshore and advertising through the internet.

A licensing regime

The introduction of a licensing regime does appear to have the support of the industry, as well as those connected to the industry, including academics and human rights organisations.[35] Under a licensing regime, companies or individuals would be required to obtain a licence for any contract that involved undertaking military and security services abroad. Deciding what activities to licence would be the responsibility of Parliament.[36] At the same time, the company in receipt of the licence would be held accountable to the government or even a parliamentary committee possibly with special powers to punish the company if any transgression of the contract took place. Similarly, the issue of accountability to the recipient state and/or its people must also be considered. How this might be achieved, however, is unclear. Whatever the outcome of accountability, because the export of military goods is already licensed and therefore accountable, it is only sensible that military services are subjected to a similar process. It is also a more flexible instrument than an outright ban.

The government would be able to consider the contract in question and its potential political and strategic impact on British interests and the interests of the recipient state and/or the people of that state. Such interests are not always the same, while a clear emphasis on the latter should be an essential component of any licensing system. A good comparison here might be with the UK and EU arms export codes. Both sets of codes require licensing decisions to give due weight to national and economic interests but expressly stipulate the need to take into account a whole series of factors in the recipient state – e.g. human rights, the potential for aggression against another state, the impact on regional security, the impact on the economy of the recipient state. Even in less controversial cases where PMCs are involved in the provision of post-conflict security for companies and aid agencies, for instance, the sheer number of such actors and the money expended on them nevertheless raise questions over whether the use of PMCs represent the best value for money for the international community and recipient states.

A further reason for emphasising the interests of recipient states and their peoples, which are not always the same, is that the UK's own military interventions have increasingly been justified by the government especially by reference to a cosmopolitanist ethic that emphasises solidarity with the oppressed. While to some extent this has merely been rhetoric, such sentiments have influenced decisions on the use of British military force abroad. It is reasonable to expect therefore, that similar values influence decisions on the licensing of UK PMCs overseas.

Finally, PMC operations are also of concern to UK departments such as the MOD, the FCO, and DFID. An additional component of any regulatory system therefore, might include some consideration of the kind of policy that agencies such as DFID should advocate to recipient states on the role and control of PMCs. DFID, for example, run a number of SSR programmes that are intended to improve transparency, democratic control and human rights standards amongst national armed forces and police forces. Increasingly, however, PMCs are taking on more security roles in developing countries, but are rarely covered by present SSR. This may be an area that an evolving policy on PMCs should address.

If the government decides to introduce a licensing regime, they should avoid drawing comparisons with the regulatory system in operation in the US, which has been proposed as a potential model.[37] There, licensing means waiting longer, but it is possible to do this commercially because of the size of the contracts. UK contracts are much smaller, therefore any delay may cause a cashflow problem. Furthermore, the majority of work undertaken by UK PMCs is for commercial customers, international organisations or foreign governments as in the case of Iraq. This customer base would expect an immediate response having entered into a contract with a PMC. A licensing regime that did not meet this requirement would place UK PMCs at a distinct disadvantage in relation to overseas PMCs, while any delay could have an adverse impact on the client's security.

There are a number of other difficulties associated with this approach. Considering such activities take place overseas, the government could find it very difficult to ensure the terms of the contract are not breached in any way. The security conditions under which a contract was issued to a PMC might change. This is extremely likely in Iraq for example if current conditions on the ground continue as they are. Any licence issued to a PMC to undertake security activities appropriate to the security environment at any time would thus have to be re-examined each time the security environment changed. This in turn could give rise to delay and work to the disadvantage of the PMC and its client.

The licensing regime may also need to take account of commercial confidentiality. It is doubtful whether a PMC would agree to a licensing regime that includes detailing commercially sensitive material on the licence that is then open to public scrutiny. Nevertheless, as the government's *Annual Report on Strategic Arms Exports* demonstrates, it is possible to have a licensing system that provides a degree of transparency, albeit very limited in this case, without compromising commercial confidentiality. Concerns of commercial confidentiality would also have to be balanced against very legitimate concerns that unaccountable elites in other countries have sometimes mortgaged the economic future of the country to purchase PMC services; a process that has been referred to as a form of imperialism by invitation. Any licensing system that did not guard against this, or at the very least provide some form of reassurance for the Parliament and the public of both the UK and the recipient state, would be highly flawed.

Such an approach might also have to take account of military security. Explaining exactly what a PMC intends to do to resolve a security predicament before acting could allow opponents the opportunity to take any necessary action to deny the PMC a successful operation. Provisions, however, could be made by the licensing authority to take account of this concern. A company employing PMCs could give assurances that it would respect human rights Conventions, the laws of the recipient state, and appropriate restrictions on the use of force. Even PMCs supporting military operations could simply provide vague details that would not compromise their operational ability. Furthermore, the inclusion of these ideas into the licensing regime would not necessarily restrict their commercial opportunities thus placing UK PMCs at a competitive disadvantage that could see them simply move offshore.

Before any licensing regime is introduced, careful consideration of its impact on the industry is vital. The security operations carried out by PMCs and some PSCs normally require them to respond immediately to a crisis, the time factor can mean the difference between the success and failure of an operation. More importantly, unlike other industries, security, or the lack of it, can mean loss of life for the employees of the clients of PMCs and PSCs. And while the industry should be subject to stringent bureaucratic overseeing to ensure professionalism and to deter cowboy operations, it is also important that the industry should not be engulfed in a bureaucratic quagmire when applying for a licence.

Whatever criteria is used to decide whether a licence for a contract should be granted or rejected will need to balance the PMC's economic interests with the interests of the UK government, the recipient state and the people of that state. Such a task will not be easy, but neither is it impossible. To assist in this task, the government could do a number of things. Contracts with pre-approved clients, for example North Atlantic Treaty Organisation (NATO), could be exempt from licensing or, as suggested by the Foreign Affairs Committee, automatic granting of licensing might be suitable for contracts carried out under the auspices of the European Union (EU), the UN or the UK government as long as firms had already been vetted.[38] This, however, will require closer involvement by the government than is presently observed.[39] Another approach would be to incorporate the automatic granting of licences into a licensing regime. The government already operates Open General Export Licences (OGELs) for the export of specified controlled goods by any company normally working in the arms industry. The OGEL removes the need for exporters to apply for an individual licence provided the shipment and destination are eligible and the conditions are met. Exporters, however, must register with the Export Control Organisation before they make use of the system. The terms of the licence remain in force until it is revoked.[40] A similar approach for controlling the sale of certain types of security services would be better for the companies and could form part of a fast track system whereby specific services on a proscribed list would be excluded from normal licensing procedures. The Foreign Affairs Committee also points out that a fast track system would save time, free administrative capacity, and would thus enable more thorough evaluation of controversial project proposals.[41] Such an approach, however, would mean that 'interested' government departments must be involved. Presently, applications for the export of restricted items, i.e. those subject to sanctions, including personal protective equipment for Iraq, are assessed by the Department of Trade and Industry (DTI), the FCO, the MOD and the Bank of England – only one of which, the FCO, is an interested department. The remainder, with only a limited interest in the subject matter, are unlikely to process applications within a commercially acceptable time frame.[42]

Another approach that would hasten the issuing of a licence might be to exclude services normally licensable but that are excluded from the licensing process because a particular project requires minimal force. The level of force needed potentially to fulfil a particular contract would determine exclusion from the process. This way it might be possible to separate out commercial contracts involving arming individuals for purposes of self-protection and government contracts where the probability of using a much greater level of force exists. Deciding a threshold would be the responsibility of the government, while the government might also exempt certain countries and regions. Here, great care would be needed to ensure the level of the threshold decided on does not cause further difficulties for a company operating in a dangerous environment. In Iraq, for example, a prime contractor specified quality 9mm weapons for close protection duties, only to find that 5.56mm or 7.62mm was essential in light of the ranges at which engagements took place. Such change might require lengthy re-application for licence revision in critical circumstances, when the provision of the new weapons themselves might take no more than a few days.[43] Consequently, any threshold set by the government may have to take account of possible changes needed to respond to a company's situation on the ground.

Alongside the introduction of a licensing regime would have to be the introduction of a system to monitor performance. Michael Grunberg, for example, maintains that included in a licensing system should be a random audit/inspection process whereby companies that have been found to have transgressed the rules under which their licence was granted would be subject to penalties, including possibly the imprisonment of their owners or executives.[44] Such a system could be paid for through the introduction of some sort of levy on the industry, while PMCs disposed to the idea should comply with the licensing conditions, freeing the government to target its resources at mercenaries, and *ad hoc* military companies that have no intention of complying with regulations.

Registration and notification

This option would in effect be a licensing system, but where the licence is automatically granted unless the government decides to withhold it. Under a registration and notification system, legislation would be introduced requiring UK PMCs bidding for military and security contracts abroad to register with the government and to inform them of those contracts they intend to tender for. This would increase the government's knowledge of the industry, while allowing the government the opportunity to intervene and prevent a company from undertaking a contract if it threatened UK interests or policy, while imposing a minimal burden on the companies.

Such an approach is a light regulatory framework and one where control is less explicit according to an ArmorGroup memorandum to the Foreign Affairs Committee.[45] ArmorGroup also pointed to the problem of notification without licence. They maintain that companies that have to notify their intention to bid for a contract will be disadvantaged because of the competitive and time-sensitive nature of the bidding process for services which may give an edge to foreign firms able to respond immediately. These services include risk analysis and planning, protection for facilities in remote and hostile environment, manned guarding, humanitarian demining, logistics, and military training and support operations. Finally, since this approach is basically a licensing system, it will be subject to

146

the same problems of enforcement, changing circumstances, confidentiality, and evasion as the options discussed earlier.

General licensing

Under this option, rather than issuing licences for individual contracts, the government would issue a general licence to a company to undertake a range of activities, while at the same time specifying what countries the companies could undertake those activities in. The licence could also impose standards that the government would expect the companies to meet. Companies for example, should not employ people with criminal records or ex-service personnel who had been dishonourably discharged.

As the Green Paper points out, on its own such a system would provide little protection for the public interest unless used in conjunction with one of the other options.[46] Using a general licence as an additional measure is the practice in the US. A general licence might also help to set an industry standard, while clients would have some of their anxiety reduced in relation to the choice of company; they would know that the firm they are employing is reputable. This system is likely to enhance the industry's image.

This approach again raises a number of important issues and questions. Issuing a general licence to companies could see the government lending credibility to companies about whose character they know nothing or who might have changed over time. The government could be perceived as supporting an operation they know little or nothing about. This, according to Christopher Beese of ArmorGroup, 'is simply unacceptable since the government should be aware of all PMC activity, whether issuing a general licence or not'.[47] On the other hand, how much change should necessitate that a company reapply? Companies could also compromise their licensing agreement unintentionally as a consequence of a modification to the structure of the business. Considering this is a service industry, operating in a fast changing security environment, such modifications are very likely as companies respond to those changes.

What criteria then would decide the eligibility of a company to be granted a general licence? The criteria could include vetting the business with respect to ownership, financial and management structures, and qualification of company personnel. Whether the criteria should include the size of the company is more contentious. Preventing, or restricting, smaller companies from bidding for large complex multifaceted contracts because the government feels they will not be able to cope will not work. As the financial scandals surrounding Enron and Arthur Anderson have demonstrated, size does not always translate into competence. The Coalition Provisional Authority (CPA) in Baghdad recently awarded a contract to Erinys International, a medium-sized British security company, to hire and train guards to protect 100 oil-related installations. The award, worth US$40m, stunned their competitors, particularly those much larger than Erinys. They questioned whether the company had the size to handle it,[48] while those who awarded the contract seem to have full trust in Erinys's ability to deliver the services.

Self-regulation

The final option is the introduction of a voluntary code of conduct. This approach, however, has proved highly contentious for a number of reasons. First, there are two schools of thought as to whether voluntary measures adequately address the growing accountability gap that has arisen out of a collective failure to assign responsibility for those human rights functions vacated by governments. Second, such a code of conduct has no legally binding enforcement mechanism for punishing companies, or that could demand extremely high standards of behaviour through sanctions that a licensing regime would demand.[49]

Some companies already subscribe to a code of conduct laid down by the ICRC for example, but only where the code is relevant to the work they are undertaking. As Beese points out, a large part of ArmorGroup's work does not involve humanitarian aid – the code deals only with the proper delivery of humanitarian aid and not to human rights – consequently the ICRC would not appreciate the company quoting the code out of context.[50] The company also subscribes to a number of other national and international codes of conduct, including the 'Voluntary Principles on Security and Human Rights' and the US[51] Foreign Corrupt Practices Act.[52]

The idea would be for companies belonging to the private military security sector to join a trade association. The government would regard membership of the Association as providing a guarantee of respectability. Under the guidance of the government, the Association, in conjunction with the companies, their clients, and NGOs would draw up a code of conduct for overseas work. Members of the Association would agree to adhere to the code.[53] Those that did not would be expelled from the Association. The immediate advantage of this approach is that there already exist three blueprints for such a code of conduct in the forms of the Geneva Convention, the Basic Principles on the Use of Force and Firearms by Law Enforcement Officials[54] and Voluntary Principles on Security and Human Rights.[55] The second document was drawn up by the UN to assist member states in their task of ensuring and promoting the proper role of law enforcement officials. The third was the initiative of the UK and US governments and aimed at the mineral extraction and oil industry and NGOs with an interest in human rights and corporate social responsibility. The intention of the initiative is to promote human rights throughout the world and to promote the constructive role business and civil society can play in advancing these goals. In each case, the trade association for PMCs and PSCs could draw on the details in the documents to help produce a code of conduct for the industry that would cover respect for human rights, respect for international law including international humanitarian law and the laws of war, respect for sovereignty, and transparency, which includes access for monitors or government representatives.[56]

The government would be able to recommend members of the Association to companies or foreign governments requiring their services, while at the same time promoting business abroad for them. The government would also not have to try to enforce what for some is unenforceable legislation or regulation, while the voluntary code would be policed by the industry.[57] External monitoring could provide a further check. In all, the option represents a simple form of regulation that would promote best practices within the industry and would enable potential clients to identify reputable PMCs.

Even so, there are a number of difficulties with this option. First, self-regulation would not meet a primary objective of regulation, to stop PMCs from damaging British interests abroad. Without the threat of legal action, all the government could do would be to watch while a company undertook a contract that plainly threatened public interest. The Association could also find itself in difficulty as a result of not knowing what exactly was happening abroad, or as a result of having to discipline associate members who fail to behave in an acceptable way. And human rights organisations would be able to con____ ____ ___ ___re accurate picture of company behaviour, par_____ _____ ____ risk areas such as Iraq and Afghanistan. A code o_ _____ ____ ___ _____ments mentioned above would detail acceptable beh_____. This ___ ____ make t easy for individuals working for human rig____ _____ _____ for monitoring PMCs to identify practices that mig____ _____ _____ ____ the moment, such individuals must rely only on th_____ _____ _____ __ an infr_ngement has occurred.

Conclusion

Current international law is inadequate to deal with PMCs that operate on the international stage. The two Conventions discussed at the start of the chapter were responses to mercenary operations that occurred during the 1960s and 1970s in post-Colonial Africa. In this respect, a new international Convention is needed to control the activities of these actors. For this to happen will require much more political will from the international community than has been demonstrated. Not only that, even if the political will was in place, there is no guarantee that such a Convention would be adopted any time soon. One only has to examine the history of the last international Convention designed to prohibit the activities of mercenaries to realise this point. The International Convention for the Recruitment, Use, Financing and Training of Mercenaries took 12 years to come into force. Adopted in 1989, the Convention did not come into actual force until October 2001. There is no reason to believe a new Convention to control PMCs will be any quicker to implement.

The only real soluti__ __ _____ __ _____, or in ce_tain cases introduce, national laws able to cc__ _____ _____ of PMCs ___ _____ tringently. In the case of ITAR, the licensing ____ _____ __ ___ _____ ____ __ ___uring the same State Department offices that ___ _____ _____ _____ __ __e same from contract to contract. It is also in_____ ____ _____ ___ __ ____ _arent as possible. To achieve this, the US go_____ _____ _____ _____ ____ ____esignating a special body to deal with contra_____ ___ ____ ____ _____ _____ to take responsibility for reporting their activi_____ __ _____. ___ ____ _____ __so be responsible for notifying Congress of c_____ _____ ____ _____ ___ __atively, the Military

149

Extraterritorial Jurisdiction Act could be changed to cover all US citizens working in the industry abroad, regardless of their clients. However, such an expansion of US criminal law may actually be unenforceable according to Singer.[58]

The Foreign Military Assistance Act also needs to be amended. By casting its net as wide as possible, tarnishing aid agencies with the same brush as mercenary outfits and PMCs, the Act could undermine their work in areas such as conflict resolution and humanitarian relief. This situation needs to be rectified as soon as possible. The Act needs to be amended so as not to blur the distinction between legitimate and peaceful humanitarian programmes from the military and security activities undertaken by mercenaries and PMCs. Not to do anything about this situation could seriously hamper the cause of peace and security, especially in Africa. Moreover, it is doubtful whether the present system, with its limited resources, will be able to handle the additional administrative workload caused by the inclusion of aid agencies. Such pressure could result in delays for agencies applying for licenses, while those South Africans working for PMCs may simply prefer to ignore the licensing process altogether, and risk a heavy fine or even imprisonment. This appears to be the situation at present in Iraq.

The British government, on the other hand, has yet to introduce any legislation to control the industry. While such a cautious approach is sensible, too much caution could also be damaging for them. The longer the government delays introducing legislation the greater the chance it will find itself reacting to international pressure to act. There may be a tendency in this situation to overreact so as to be seen to be doing something positive. The outcome could be the introduction of a rushed licensing system that has not been carefully thought out, and which might result in more problems for the government to contend with than if no system had been introduced at all. Also, because of the Green Paper, the government is ideally placed to take the lead in promoting a European wide Act to control the industry. As yet, no other European government has examined ways of doing this.

Finally, the absence of any suitable legal mechanism to control the industry cannot be allowed to continue. The military/security services undertaken by PMCs should be of vital concern to the public. In the long term, international law must respond to the new set of challenges posed by PMCs. In the short term, however, there is an urgent need for governments to overhaul the national laws relating to this area, or as in the case of some countries to introduce new set of laws that are able to establish a clear set of standards for PMCs to meet and, if they do not, are able to punish them for failing to do so.

CONCLUSION

Whether we agree with them or not, PMCs and PSCs have established themselves as important actors on the international stage. As with any international system, the provision of private security is closely related to the structure of the security sector as a whole. The key factors that define this context in today's international system are military downsizing, technology, monetary efficiency and political expediency. Nor are these actors likely to disappear very soon. This would require a change to the nature of the international system. Even then, certain parts of the industry, particularly information technology companies that service advanced weapons systems, would in all probability remain.

In every respect, private security is now a global phenomenon that is set to grow. Such growth, however, is likely to be confined to tier 2 and 3 of Singer's typology,[1] or as illustrated in Chapter 1, the bottom two quadrants of the axes. These areas are concerned with the provision of military advice and training, security services, logistics, intelligence, technical support, supply and transportation. At the same time, companies that want to provide direct military assistance, as in the case of EO during the Angolan and Sierra Leone Civil War, may find this market limited in the future. Nor is it likely to be an easy ride for the industry. Companies will continue to experience problems while international organisations, governments and civil societies adjust to this new reality. There is still a lot of hostility towards PMCs, especially from sectors of the media, which still refer to them as mercenaries. Such labels are given without much thought as to the real nature of PMCs, and tend to reflect past perceptions of private security that are, quite frankly, outdated. Nor will it all go the industry's way. While many PMCs are cashing in on the insecurity in Iraq, as a result of the Bush Administration's determination to keep US troop level in the country down, outsourcing this way is problematic. Instead of deploying more troops, the Administration appears to be handing over more and more security responsibility to the private sector. Unfortunately, the industry is still at an early stage of its development and as such does not really know its functional limits. Some UK PMCs have already expressed concern over what they see as a hiving off of offensive roles to companies that are neither equipped nor qualified to take them on. The 'yes we can do it' attitude is starting to take hold in companies, whilst little or no regard is given to the consequences if they get it wrong, and is a potential problem for governments that will have to sort out the mess. In this respect, governments need to be aware of the functional limitations of the industry so as not to outsource roles that are outside the functional capabilities of companies.

The book itself makes a number of claims about the industry that are listed below and explained in greater detail in the following paragraphs:

1 That PMCs are a new actor, with a different nature to previous examples of private force;
2 That this newness is increasingly being recognised;
3 That PMCs are having an increasing impact upon international security;
4 That the impact can be positive for stability and security;
5 That governments have yet to find a satisfactory legislative solution to this global phenomenon.

As mentioned above, PMCs are new actors to international security. They also represent a different type of private violence from previous examples. There is no doubt that US PMCs have their roots in the corporate sector, while the history of UK PMCs is more controversial. Their roots can be traced back to the mercenary operations that took place during the 1960s and 1970s. However, a split occurred in the industry around the mid-1970s. Companies such as Control Risks Group were able to exploit the growing demand for legitimate security by the corporate world, while mercenaries continued to operate at the very edge of international law. This newness is increasingly being recognised by international organisations, governments, and civil society. Even Dr Shameem, the new UN Special Rapporteur for Mercenaries, agrees that the definition of mercenary does not necessarily apply to all security companies, implying, therefore, that they are different to mercenaries.[2] The process, however, is slow moving, apart from in the US where the legitimacy of PSCs has been recognised for some time, and as a consequence the companies are now actively involved in supporting US foreign policy. The issue of recognition needs to be resolved sooner than later, especially now that more and more international organisations, governments, multinational corporations, and even charities, are turning to PMCs for security or advice on security.

If the war in Iraq has demonstrated one thing, it is the extent to which some governments now rely on the services PMCs offer. Such reliance is not confined to Iraq, nor is it evenly spread. While UK PMCs are still struggling to be accepted as full members of the development–security community, US PMCs are already undertaking security sector reform programmes for the US government throughout Africa and other parts of the developing world. By training state militaries and supporting humanitarian operations US PMCs are already having a positive impact on developing counties while strengthening international peace and security. However, as just mentioned, PMCs have functional limitations, and to go beyond their limitations will surely result in companies facing serious problems. In this respect, there will always be operations beyond the physical capabilities and experience of PMCs.

PMCs are potentially a very useful instrument for governments facing an array of socio/political and security dilemmas, humanitarian organisations confronting human disasters, and international organisations tasked with operating in volatile regions of the world. PMCs offer a cost-effective way to manage security for these groups by using highly trained and proficient former soldiers, typically, but not always, from First World armies, with many years experience of operating in conflict zones and trained in a range of skills. Such a resource used properly can make a significant difference in a conflict environment either by using private contractors to advise on security arrangements or taking direct action to prevent a security problem from occurring or worsening.

Contextualising the position of UK and US PMCs in the international security environment

While the book is about the PMC industry in general, most of what has been written is concerned with the activities of UK and US PMCs since UK and US companies dominate the market. Neither will the role of PMCs change too drastically over the next few years. Iraq may have whetted their appetite for bigger operations, but it is more likely they will continue to operate around their core competencies. In this respect, we are not going to see the developing world swamped by PMCs imitating the type of operations that EO undertook back in the mid-1990s, unless governments demand it, since there are only a handful of such companies that possess both the expertise and personnel to even contemplate such operations, let alone succeed.

Some experts on the subject, notably Jim Hooper,[3] even question whether EO's success can actually be copied. Hooper argues that EO's success is directly attributed to their operators being former Permanent Force cadres of the SADF, with extensive experience in joint conventional/special forces combat operations in Angola. As such, they knew of each other from 20 years of fighting the Front Line States, while they were also intimately familiar with the terrain, weather, and infrastructure of Angola, a consequence of conducting covert and deep penetration operations. Furthermore, having fought the FAPLA/FAA[4] forces of their MPLA client, they were well aware of their new charges' strengths, weaknesses, and mindsets. Conversely, they were even more familiar with the strengths, weaknesses and mindsets of UNITA, many of whose senior officers they knew personally. Finally, the timing and political atmosphere of the Angolan and Sierra Leone contracts was impeccable. Both these countries were faced with opponents whose human rights records and unwillingness to compromise were already roundly criticised by the international community, while the ideological and geopolitical struggle between the US and USSR was history, thus precluding competing interference by either superpower.

The larger commercial security companies, on the other hand, prefer not to involve themselves in this area of security because of the potential damage to their business if something goes wrong. The type of services offered by these companies will continue to expand, reducing the need to take on more controversial

153

operations that were the earmark of EO. The type of services these companies provide include protecting company assets and company employees working in dangerous regions, humanitarian support, retraining state militaries and law enforcement agencies, and low intensity policing. As such, their role in the field of security is unlikely to resemble the type of state security undertaken by EO. Some of these companies do not even like arming their employees, and when they do it is only for reasons of self-protection and the protection of their client and not to carry out offensive operations.

It is clear that major differences exist between UK and US companies in terms of their roles and the nature of their relationship to the foreign policy of their country. In each case, however, the companies are also the consequence of incapacity or unwillingness on the part of the military to deliver the level of security that we have expected from them in the past.

The nature of the relationship between UK PMCs with the government is still without consideration to wider implications. The industry still organises itself around informal structures, including old boy networks, while adopting the nod and wink business practice approach, instead of using formal structures able to promote best business practices to organise the industry. In this respect, there appears to be a lack of political will on the part of the government to do anything about this by using regulation, while the probable cause of the lack of political will was the 'arms to Africa' debacle. The debacle has undeniably impacted deeply on the relationship between the government and PMCs, forcing the government to maintain its distance to safeguard its image among voters. How long the government can maintain this position is unclear.

This position stands in marked contrast to the position of US PMCs that are seen as utilising a national resource in the form of the retired military community,[5] while at the same time being an integral part of the US defence sector and increasingly used as a foreign policy resource. Many US PMCs actually work alongside the armed forces. The US PMC market is also better organised structurally to meet both client and operational demands, while the relationship the industry has with the government is formally structured. The industry has been able to integrate itself far better into the US defence sector and foreign policy environment than their UK counterparts. As such, the industry has been able to meet the demands now being placed on it by government agencies in an efficient and cost-effective way.

The most notable difference between the UK and US sectors is their size. Multinational corporations tend to dominate the US PMC market, while UK PMCs are much smaller, the largest companies employing at most several thousand employees globally. Even so, each still undertakes the same type of security activities, which could lead to greater international integration within the security market. UK PMCs could thus see the sharp end of the market they have dominated for so long increasingly threatened by US PMCs competitors. Whether this would place them at a commercial disadvantage is not clear. With US dominance of the global security market likely to increase following 11 September, UK companies will, however, find it increasingly harder to compete on equal terms unless support from the UK government is forthcoming.

As we know, the industry's share of the global security market has risen in the last decade. While such a development should not overly worry the armed forces around the world, it is possible that in the near future state militaries will become a recruiting ground for PMCs. This is particularly so with Special Forces personnel who today are being siphoned off to meet demand for private security in countries such as Iraq and Afghanistan. This problem is especially worrying for the UK government because of the importance the commercial sector placed on hiring British Special Forces, while at the same time being able to offer much more lucrative financial rewards as an inducement to move across to the private sector. It is very possible that in the not too distant future, if it is not already happening, regiments, such as the SAS, will become a training ground for soldiers whose main interest is in commercial security, but see attaining the coveted winged badge as a way of improving their employment possibilities.

Finally, both governments should take the close working relationship between companies from each country very seriously. The transnational nature of the business could result in either the UK or US governments facing legal challenges as a consequence of the action of one of their citizens operating for a PMC registered on the other side of the Atlantic. This problem is not confined to the governments of the UK and US. Any government that allows its citizens to operate for a PMC registered in another country could face the same problem. US PMCs, for example, are now hiring former Polish commandos to protect Baghdad airport,[6] while former French and South African soldiers are on a database operated by Northbridge, an Anglo–American PMC.[7] The problem appears more acute in the UK and US since PMCs in these two countries do appear to be drawing closer together. This area needs further research as soon as possible to avoid legal complications for both governments.

Managing the industry

The most contentious area of debate at present is among human rights advocates, who see PMCs as posing a direct threat to human rights, and those who support the use of PMCs who argue that such a threat has yet to materialise and probably will not. It is true that the mercenaries who fought in the Congo, Angola, and other wars of decolonisation back in the 1960s and 1970s, did not generally give any real consideration to the laws of war, except maybe in individual cases. It is therefore not surprising that human rights groups are concerned about PMCs acting in the same way. The roots of UK PMCs are, after all, firmly fixed in the colonial wars fought during the 1960s and 1970s when such violations were common. But making a moral judgement on the intention of an employee of a PMC to act illegally based on a perception that is rooted in a different era, where the employee is different, and encompassing a very different set of social, political and economic realities, is problematic.

The private security contractor working for a PMC today is quite different from the classic mercenary who fought in Africa's wars of decolonisation. Much has been made of the threat PMCs pose to the democratic process in countries plagued by civil war, and the introduction of military force into civil wars by classic mercenaries has certainly posed such a threat in the past. There is, however, no substantive evidence to suggest PMCs have attempted to do the same. Companies have, in the past, been very careful about engaging in civil wars, realising their presence can make matters worse, while in the case of Angola and Sierra Leone, EO's presence in these two countries actually strengthened the democratic process.

There are other reasons why PMCs need to be regulated and monitored, and why regulating them should be a priority for all governments. First, by monitoring them, governments could ensure they are held accountable both for their actions as well as the actions of their employees, while regulating them will help to make them more transparent. Here governments should be able to regulate what the companies must disclose about their business and what they can keep confidential. After all, all companies are entitled to a level of confidentiality in the conduct of their business. Full engagement with governments should prevent them from shifting between legal and illegal pursuits and should also guarantee they do not undermine the sovereignty of states.

Finally, PMCs do abide by international standards relating to the laws of armed conflict, while advocates of PMCs also point to this. Thus, the threat posed by PMCs to human rights in the main has not materialised. Why? Much of what is written on the subject argues that because PMCs do not operate within a formal structure that can accommodate sanctions against individual persons who break the laws of war, such persons will act beyond the laws of war because there is no formal structure to enforce the laws of war such as the court martial system in which state militaries operate. The conclusions drawn here suggest otherwise. Through a matrix of processes and relationships involving, in particular, socialisation, the constraints of commercial contacts, and social awareness, the behaviour of private contractors working for PMCs can be controlled, though one must accept that weaknesses in the system, which should be recognised, are something missed by the checks and balances introduced to stop a violation from occurring. The same problem applies to state militaries. As such, more research into this area is urgently needed, especially now states are considering regulating PMCs.

Regulating the industry

It is unlikely that international law to regulate the activities of PMCs will be introduced in the immediate or not too distant future. An examination of previous attempts to use international law to control the activities of classic mercenaries highlights this point. In 1989, the UN General Assembly adopted and opened for signature and ratification the International Convention against the Recruitment, Use, Financing and Training of Mercenaries.[8] The Convention was introduced 14 years after the mercenary debacle in Angola, and was intended to stop such a debacle happening again. The international community had to wait until 2001 before enough signatories actually ratified the Convention making it law.[9] To suggest, therefore, that the international community can introduce international legislation to regulate PMCs in the next couple of years is simply not realistic.

US and South African PMCs are already regulated at the national level; the former through ITAR and the latter through the Foreign Military Assistance Act. Briefly, the export and import of all US defence articles and defence services, including those supplied by PMCs are controlled through the Arms Export Control Act. Under the Act, only the President is authorised to issue licences for the exporting and importing of defence articles and services.[10] The system, however, has no suitable oversight mechanism in place, while this will need to be addressed if public confidence in the industry is to grow. Neither has the South African Foreign Military Assistance Act proved very effective. While there have been a number of successful prosecutions under the Act, the Act does not seem to have dampened demand for former South African soldiers in war zones around the world. Nor is it clear whether strengthening the Act further will stop the demand for their services. Therefore, given the situation and the inadequacy of the Act, it needs to be reviewed if the South African government is to have any chance of controlling the industry.

The situation in other countries is again different to that in the US and South Africa. Apart from Britain, which is still considering legislation, most other countries have not yet thought about it. But passing national legislation should be a priority for governments that allow PMCs to operate from their territory; a possible exception being EU countries, which could consider an EU regulatory system instead. Before any regulatory system is introduced, governments will need to consider what is the best way to regulate the industry. To suggest, for example, that the UK adopts a system similar to, if not the same as, the system used in the US, will result in a debacle. The reason for this is quite simple. While both industries supply the same services, they approach the business in a different way. What are needed are national regulatory systems that are responsive to the business practices of different countries and that also respond to domestic business practises.

Ultimately, however, the hardest problem will be achieving the right legal balance between the wider interests of the general public and the commercial interests of PMCs. There is also the impact of regulation on developing countries to consider. Removing PMCs from the international security arena will do little to dampen demand for their services, while their removal could make matters worse for local government and national corporations that have come to rely on their services. Such a security vacuum is likely to be filled by the very individuals regulation of the industry is supposed to eradicate: mercenaries.

Final comments

There is nothing unique about the privatisation of violence. A very quick inspection of history reinforces this point. Mercenary companies in Europe were used in war right up to the start of the nineteenth century, while classic mercenaries operated in the Congo in the 1960s and Angola in the early 1970s. Thomson is right to be highly sceptical about the utility of governmental monopolies on the use of force to promote security.[11] What is unique today about private security is the manifestation it has taken. PMCs are very different to forms of private violence that have emerged in the past. But our perception of PMCs in the UK is still influenced by the events that took place in the Congo and especially Angola, while such tensions and contradictions finally surfaced with the 'arms to Africa' affair. The affair also represents a turning point for PMCs in that after the affair, there was recognition on the part of Parliament that PMCs were different to mercenaries and that they could contribute to international security as long as they were regulated. This process is ongoing. The US on the other hand has avoided any characterisations of PMCs as representing anything more than mercenaries hiding behind a corporate veneer. There is not only greater acceptance in the US of PMCs, the US government actively use them to promote a range of government initiatives.

Outsourcing to PMCs would see them taking on roles that complement the role of the state military, but do not replace it. Ultimately, a situation could emerge whereby the state military only undertakes tasks associated with its core combat competencies, while the other tasks are undertaken by civilian contractors. Thus, the military represents the spearhead and civilian contractors the shaft of the spear. Each component moreover needs the other to function, while it is intended that such a division of labour will increase efficiency. This situation is not likely to occur in the immediate future, but we should not dismiss it. As the war in Iraq has shown, not only are military forces becoming more reliant on the private sector to operate, civilian contractors are getting closer to the front line.

The long-term future of the industry appears positive. While many people still object to the security services offered by PMCs, the trend towards outsourcing is increasing. This strongly suggests that such security services will remain a part of international security for some time to come. At the same time, clients lack the necessary knowledge of the industry to question effectively the competency of companies, let alone ensure their actions abide by international law when conducting security operations on their behalf. The inability of clients to make informed decisions in this area is worrying. We could see those in charge of multinational corporations facing criminal charges as a consequence of the actions of the PMC. Such problems also place limits on the credentials of PMCs. A properly licensed industry is thus essential if both these problems are to be addressed effectively. With the licensing of PMCs, clients will be able to make an informed decision about which PMC they want working for them. At present, clients are often in the dark about the level of expertise and professional standards exhibited by individual PMCs.

Today, PMCs operate openly for major clients, including government agencies and multinational corporations. But there is still public distrust towards the industry, especially in the UK. The industry is caught between two sets of opposing interests represented either as legal or social forces. Legal forces have dramatically transformed the legal personality of PMCs over the last decade. This has been necessary for the industry to prosper financially, but also, as the industry has grown, the need to conduct its business within legal boundaries has become paramount. At the same time, the industry is struggling to change its image in the eyes of the general public from that of the classic mercenaries of the 1960s and 1970s to professional security consultants. This has not been an easy task and is still ongoing. As with the British mercantile companies that sought an intermediate approach to commerce that would satisfy the diffusion of power between the monarch and Parliament, the UK PMC could yet transform itself by taking an intermediate approach that sought to satisfy the commercial interest of the companies and wider national interests of Parliament. Needless to say, to do so would require the help of the government through the introduction of a properly thought-out licensing regime able to set a code of practice for PMCs to abide by.

Finally, our knowledge of PMCs is still fairly limited when compared, for example, to what we know about state militaries. But then, PMCs only emerged on the international stage just over a decade ago. Clearly there is huge scope for further research into PMCs. At present the field appears bare. Researchers appear reluctant to engage with PMCs because of the mercenary label that has been attached to them. As this book shows, this label does not fit the majority of PMCs now working in the field of security, while the distinction between PMC and mercenary will increase over the next decade. In the meantime academics, government officials, and the media need to engage with the industry if for no other reason than to learn as much about it as possible. Only then will the general public start to feel comfortable with the idea of using PMCs to undertake security tasks for legitimate actors. It is hoped this book has gone some way to addressing this by furthering our knowledge of the subject.

NOTES

INTRODUCTION

1 The US has an expenditure on private security reported in 1999 as at US$50 billion, which is larger than the defence budget of every other NATO country added together. R. Mandel, Armies without States: the Privatisation of Security, London: Lynne Rienner Publishers, 2002, p. 8.

2 P. Singer, Corporate Warriors: The Rise of the Privatised Military Industry, London: Cornell University Press, 2003, p. 80.

3 Foreign Affairs Committee, Private Military Companies, London: The Stationery Office, 23 July 2002, p. 6, paragraph 6. Also see Foreign and Commonwealth Office, Private Military Companies: Options for Regulation, London: The Stationery Office, 2002, p. 8, paragraph 17.

4 Foreign Affairs Committee, op. cit., p. 6, paragraph 6.

5 C. Beese, 'Private Security Companies'. Email, 13 November 2003.

6 Mandel, op. cit., p. 4.

7 D. Isenberg, 'Security for Sales', in Asia Times, 14 August 2003.

8 The International Consortium of Investigative Journalists, The Business of War, Washington, DC: The Center for Public Integrity, 2002, p. 2.

9 Singer, op. cit., pp. 11–17.

10 ArmorGroup Brochure, Single Source Solutions for Global Security Risk Management, London: ArmorGroup, 2001.

11 MPRI, http://www.mpri.com (accessed 18 June 2002).

12 Singer, op. cit., p. 11.

13 O. Morgan, 'Soldiers of Fortune Hit the Jackpot', The Observer, 26 October 2003.

14 Foreign and Commonwealth Office, op. cit., p. 37.

15 Annex A of the Green Paper details mercenary and PMC activities in Africa from the 1950s to the 1990s.

16 See section on definitions below for an explanation detailing the differences between PMCs and PSCs.

17 For a detailed account of the services provided by PMCs and PSCs see D. Shearer, 'Private Armies and Military Intervention', in Adelphi Paper 316, 1998, pp. 25–6.

18 Singer, op. cit., p. 230.

19 A-F. Musah and J. Fayemi (eds), Mercenaries, London: Pluto Press, 2000, p. 14.

20 Mandel, op. cit., p. 158.

21 Musah and Fayemi, op. cit.

22 Shearer, op. cit., p.22.

23 S. Shameem, Interview. Online at http:// www.abc.net.au/ra/pacbeat/stories/m926500.asx (accessed 8 October 2004).

1 A TYPOLOGY OF PRIVATE MILITARY/SECURITY COMPANIES

1 This term was first introduced by Mr Lafras Luitingh of Sandline International during a conference on 24 June 2002 at Birmingham University to discuss the government's Green Paper on options for regulating PMCs.
2 D. Lilly, 'Regulating Private Military Companies: The Need for a Multidimensional Approach', London: *International Alert*, 2002, p. 2.
3 D. Shearer, 'Private Armies and Military Intervention', *Adelphi Paper 316*, 1998, pp. 25–6.
4 Lilly, op. cit., p. 2.
5 Interview with Lt Gen. Ed Soyster (Retd), 10 September 2002.
6 T. Dunne, 'Realism', in Baylis J. and Smith S. (eds), *The Globalisation of World Politics*, Oxford: Oxford University Press, 1997, p. 122.
7 The Foreign and Commonwealth Office, *Private Military Companies: Options for Regulation*, London: The Stationery Office, 2002, p. 11.
8 M. Kaldor, 'Introduction', M. Kaldor, and B. Vashee (eds), *New Wars*, London: Pinter, 1997.
9 Nic van den Berg served as the operations manager for Sandline International.
10 Interview with Nic van den Berg, 24 June 2002.
11 N. Shachtman, *Hired Guns*. Online. Available at http:// www.techcentralstation.com (accessed 24 July 2003). The article discusses an offer put forward by a consortium of US PMCs to bring an end to the war in the Congo.
12 J. Hooper, *Bloodsong!*, London: Collins, 2002, p. 47.
13 Shearer, op. cit., p. 23.
14 Sandline International, 'Private Military Companies: Independent or Regulated?'. Online. Available at http://www.sandline.com/site/index.html (accessed 11 July 2002).
15 Ibid.
16 C. Beyani and D. Lilly, *Regulating Private Military Companies*, London: International Alert, 2001, p. 16.
17 T. Spicer, *An Unorthodox Soldier*, London: Mainstream Publishing, 1999, p.165.
18 P. Singer, 'Corporate Warriors: The Rise of the Private Military Industry and its Ramifications for International Security', *International Security* Vol. 26, No. 3, 2002, pp. 198–200.
19 MPRI, http://www.mpri.com (accessed 18 June 2002).
20 An issue discussed at a conference at Birmingham University on the options for regulating PMCs. Conference on 'Options for Regulating PMCs', Birmingham University, 24 June 2002.
21 Human rights activists see relocation as a way of avoiding regulations. Beyani and Lilly, op. cit., p. 17.
22 A-F. Musah and J. Fayemi, *Mercenaries*, London: Pluto Press, 2000, p. 1.
23 Interview with Lt Gen. Ed Soyster (Retd), 10 September 2002.
24 As discussed later, the use of deadly force by companies such as ArmorGroup is as a final countermeasure against a potentially lethal attack directed against the company's personnel or the client. Interview with Nigel Woof, 30 November 2001.
25 MPRI will only work for legitimate governments, and not business corporations or other private interests. Interview with Lt Gen. Ed Soyster (Retd), 10 September 2002.
26 MPRI, http://www.mpri.com/channels/int_overview.html (accessed 12 June 2002).
27 Shearer, op. cit., p. 34.
28 Ibid., p. 36.
29 Beyani and Lilly, op. cit., p. 18.
30 ArmorGroup, *Regulation: An ArmorGroup Perspective on Private Security Companies*, London: ArmorGroup, 2004, p. 1.

31 Beyani and Lilly, op. cit., p. 18.
32 D. Shearer, 'Private Military Companies and Challenges for the Future', *Cambridge Review of International Relations*, Vol. XIII, No. 1, 1999, pp. 85–6.
33 A. Vines, 'Mercenaries and the Privatisation of Security in Africa', *The Privatisation of Security in Africa*, South Africa: South African Institute of International Affairs, 1999, p. 73.
34 Interview with Michael Grunberg, 29 November 2001.
35 Inside Story, 'BP's Secret Soldiers', *The Guardian*, 4 July 1997.
36 Vines, op cit., p. 71.
37 In January 1998, and at the request of BP Exploration Colombia, the Public Prosecutor's Office of Colombia completed a 14-month preliminary investigation of the accusations that BP was involved in human rights violations. In that investigation, the Public Prosecutor's office found no substance to the accusations, and as such no grounds for opening a formal investigation. Interview with Andres Penate, Director International Affairs BP, 11 September 2002. Also see report by UN Special Rapporteur on mercenaries. E. Ballesteros, *Use of Mercenaries as Means of Violating Human Rights and Impeding the Exercise of the Right of People to Self Determination*, A/54/326, Geneva, The Office of the High Commissioner for Human Rights, 1999, p. 19.
38 Armour Holdings Inc. acquired DSL in 1997 to form the basis of its services division and several smaller acquisitions were added to it between 1998 and 2001. The division was renamed ArmorGroup.
39 Interview with Nigel Woof, 30 November 2001.
40 Beyani and Lilly, op. cit., p. 18.
41 ArmorGroup, http://www.armor-mine-action.com/ (accessed 15 July 2002).
42 Shearer, op. cit., p. 85.
43 Group 4 Falck A/S's security business merged with Securicor Plc to form Group 4 Securicor in July 2004.
44 Group 4 Securicor, http://www.group4securicor.com (accessed 24 January 2005).
45 R. Sutton, Group Human Resources Director, Group 4 Securicor, interview, 30 July 2002
46 K. O'Brien, 'Military–Advisory Groups and African Security: Privatised Peacekeeping', in *International Peacekeeping*, London: Frank Cass, Vol. 5, 1998, pp. 78–105.
47 A. Mockler, *The New Mercenaries*, London: Corgi Books, 1986, pp. 35–6.
48 T. Spicer, 'Interview', *Cambridge Review of International Affairs*, Cambridge: Vol. XIII, No. 1, 1999, p. 165.
49 Article 47(2), Protocol I Additional to the Geneva Convention 1949.
50 Idealists can be found on either side in a conflict. Those British mercenaries who fought in Angola included Nick Hall. Hall claimed he was fighting not for money, but against communism, even if it meant fighting without pay in a distant country far from home. C. Dempster and D. Tomkins, *Firepower*, London: Corgi, 1978, p. 59. Hall's political views were in fact extreme rightwing. Mockler, op. cit., p. 220.
51 For a detailed discussion on the legal position of mercenaries in international law, see F.J. Hampson, 'Mercenaries: Diagnosis before Prescription', *Netherlands Yearbook of International Law*, Vol. 22, No. 3, Netherlands: Martinus Nijhoff Publishers, 1991, pp. 28–30.
52 A-F. Musah, 'A Country Under Siege: State Decay and Corporate Military Intervention in Sierra Leone', in Musah and Fayemi, op. cit., p. 88.
53 Ibid., p. 88.
54 K. O'Brien, 'Military–Advisory Groups and African Security: Privatised Peacekeeping', *International Peacekeeping*, London: Frank Cass, Vol. 5, 1998, pp. 78–105.
55 Mockler, op. cit., pp. 35–6.

56 For a detailed discussion that seeks to establish a clear difference between the classical view of mercenaries as hired guns and the more recent, business oriented, phenomenon of PMCs and PSCs, see G. Cleaver, 'Subcontracting military power: The privatisation of security in contemporary Sub-Saharan Africa', *Crime, Law & Social Change*, Netherlands: Kluwer Academic Publishers, Vol. 33, 2000, pp. 131–49.

57 K. Silverstein, *Private Warriors*, London: Verso, 2000, p. 169.

58 No. 15 of the Regulation of Foreign Military Assistance Act 1998.

59 Interview with Nic van den Berg, 24 June 2002.

60 D. Isenberg, 'A Profile of Today's Private Sector Corporate Mercenary Firms', *Center for Defense Information*, Washington, DC: Centre for Defense Information, 1997.

61 The author accepts that not all state militaries follow the traditional Westphalia model. Occasionally a state military is used to secure private objectives, locating A in the top left-hand box, instead of the top right-hand box.

62 While state militaries are the ultimate fighting force, irregular forces can challenge them, for example. M.V. Crevald, *On Future War*, London: Brassey's, 1991.

63 Sandline International, http://www.sandline.com/company/index.html (accessed 30 July 2002).

64 For the details of the weapons Sandline International purchased see A. Vines, 'Mercenaries, Human Rights and Legality', in Musah and Fayemi, op. cit., pp. 191–2.

65 House of Commons Foreign Affairs Committee, *Private Military Companies*, London: The Stationery Office, 23 July 2002.

66 J. Tepperman, 'Out of Service', *New Republic*, 13 January 2003.

67 MPRI, http://www.mpri.com/channels/mission.html (accessed 24 March 2005).

68 Interview with Lt Gen. Ed Soyster (Retd), 10 September 2002.

69 MPRI, http://www.mpri.com/subchannels/int_africa.html (accessed 2 August 2002).

70 House of Commons Foreign Affairs Committee, Evidence 82, 23 July 2002.

71 DynCorp International LLC provides international services to support the goals and objectives of its worldwide customer base. These services include technical and logistical services. They include aviation services, marine services, logistical support services, and personal and physical security services. It is this division that would have been responsible for supplying former police officers to the international police force in Bosnia. DynCorp International, http://www.dyncorp.com/areas/index.htm (accessed 9 January 2003).

72 DynCorp International, http://www.dyncorp.com/areas/index.htm (accessed 24 March 2005).

73 Corpwatch, http://www.corpwatch.org/issues/PID.jsp?articleid=672 (accessed 5 August 2002).

74 J. O. Tamayo, 'US Civilians on Frontlines of Drug War in Colombia', *Knight Rider Newspaper*, 26 February 2001.

75 While the official name of the company has changed to ArmorGroup International PLC, the company is still referred to as ArmorGroup.

76 ArmorGroup Brochure, *ArmorGroup: Single Source Solutions for Global Security Risk Management*, London: ArmorGroup, 2001.

77 Interview with Nigel Woof, 30 November 2001.

78 Interview with Christopher Grose, 11 May 2004.

79 I. Traynor, 'The Privatisation of Warfare', *The Guardian*, 10 December 2003.

80 Interview with John Holmes, Director, Erinys International, 8 July 2004.

81 Interview with John Holmes, 8 July 2004.

82 The company does have a subsidiary company concerned with high-risk security contracts of the sort normally undertaken by companies such as ArmorGroup. In this instance, since it is a separate company, the position of Group 4 Securicor remains unchanged on the axes.

83 Interview with Richard Sutton, Group Human Resources, Director, Group 4 Securicor, 30 July 2002.
84 Group 4 Securicor, http://www.group4falckgsl.com/ (accessed 7 August 2002).
85 Interview with Richard Sutton, 30 July 2002.
86 Ibid.
87 Ibid.
88 Interview with Michael Grunberg, 24 June 2002.
89 For a detailed list of companies see Musah and Fayemi, op. cit., pp. 265–74.
90 A.-F. Musah and J. Fayemi, 'Africa in Search of Security: Mercenaries and Conflicts, an Overview', in Musah and Fayemi, op. cit., p. 61.
91 Lifeguard was awarded the contract to protect the dam from attacks by the RUF. For details about Lifeguard's role in Sierra Leone see, Legg Inquiry: Note of the Hearing with Lt Col. Tim Spicer OBE (Sandline International) held at 2.00 pm on 24 June 1998, or visit Sierra Leone, http://www.sierra-leone.org/slnews1096.html (accessed 14 May 2003).
92 K. Pech, 'The Hand of War', Musah and Fayemi, op. cit., pp. 140–4.
93 Beyani and Lilly, op. cit.
94 Shearer, op. cit., pp. 34–7.
95 Interview with Lt Gen. Ed Soyster (Retd), CDI, http://www.cdi.org/adm/1113/shoyster.html, 19 March 2001.
96 This point was made very clear in an interview with Richard Sutton, 30 July 2002.

2 AN HISTORICAL OVERVIEW OF PRIVATE VIOLENCE

1 J. Thomson, *Mercenaries, Pirates, and Sovereigns*, Princeton, NJ: Princeton University Press, 1994, p. 21.
2 Ibid., p. 41.
3 W. McNeill, *The Pursuit of Power*, Oxford: Basil Blackwell Publishers Ltd, 1984, p. 66.
4 Ibid., p. 66.
5 P. Kennedy, *The Rise and Fall of the Great Powers*, London: Fontana Press, 1988, p. 25.
6 McNeill, op. cit., p. 74.
7 Such social and economic changes that occurred in Northern Italy, weakening the knights' supremacy in war, was followed by a similar set of changes taking place in the Low Countries, where a secondary commercial centre was being established by the twelfth century. Overland portage routes that allowed for the exchange of artisan activity eventually linked both commercial centres. As trade between these two centres increased, the wealth generated started to alter the social balance in each centre in favour of the merchant-capitalist. As a result of these changes, the importance of the knight and his relationship to the rural community was undermined. Ibid., p. 66.
8 M. Howard, *War in European History*, Oxford: Oxford University Press, 1977, p. 20.
9 Ibid., pp. 21–2.
10 For an account of the military revolution that occurred between 1560–1660 see, M. Roberts, 'The Military Revolution, 1560–1660', in C. Rogers (ed.), *The Military Revolution Debate*, Oxford: Westview Press Inc, 1995.
11 Ibid., p. 15.
12 Ibid., p. 29.
13 W. E. Hall, *A Treatise on International Law*, Oxford: Clarendon Press, 8th edition, 1924, pp. 620–1.
14 F. Sherry, *Raiders and Rebels: The Golden Age of Piracy*, New York, NY: Hearst Marine Books, 1986, p. 57.
15 Thomson, op. cit., p. 70.
16 Ibid., p. 23.

17 Ibid., p. 23.
18 Ibid., p. 24.
19 F. R. Stark, *The Abolition of Privateering and the Declaration of Paris*, New York, NY: Columbia University, 1987, pp. 82–6.
20 Thomson, op. cit., p. 24.
21 P. Crowhurst, *The Defence of British Trade, 1689–1815*, Folkestone: Dawson, 1977, pp. 15–17.
22 G. Symcox, *The Crisis of French Sea Power 1688–1697*, The Hague: Martinus Nijhoff, 1974, pp. 221–33.
23 Thomson, op. cit., p. 26.
24 Ibid., p. 32.
25 Ibid., p. 33.
26 Ibid., p. 34.
27 E.L.J. Coornaert, 'European Economic Institutions and the New World: The Chartered Companies', in E.E. Rich and C.H. Wilson (eds), *The Cambridge Economic History of Europe*, Cambridge: Cambridge University Press, Vol. 4, 1967, p. 247.
28 H. Furber, 'Rival Empires of Trade in the Orient, 1600–1800', in B. Shafter, *Europe and the World in the Age of Expansion*, Oxford: Oxford University Press, Vol. 2, 1976, p. 91.
29 C.R. Boxer, *The Dutch Seaborne Empire 1600–1800*, London: Hutchinson & Co., 1972, p. 25.
30 Ibid., p. 24.
31 Thomson, op. cit., p. 37.
32 Ibid., p. 38.
33 Furber, op. cit., pp. 93–4.
34 Thomson, op. cit., p. 39.
35 Furber, op. cit., p. 40.
36 Ibid., pp. 46–7 and 63.
37 Thomson, op. cit., p. 40.
38 Ibid., pp. 40–1.
39 Furber, op. cit., pp. 43–4.
40 D. Mackay, *The Honourable Company: A History of the Hudson Bay Company*, New York, NY: Bobbs Merill, 1936, pp. 134–5.
41 Thomson, op. cit., p. 61.
42 Ibid., p. 61.
43 Roberts, op. cit., p. 15.
44 Ibid., p. 15.
45 Ibid., p. 15.
46 For a brief account of the impact of the industrialisation of war, see Kennedy, op. cit., pp. 183–248.
47 Horses and mules were still vital to armies well into the twentieth century. But, the introduction of steamships and railroads reduced their significance. A. Beevor, *Stalingrad*, London: Penguin, 1999, p. 227.
48 Even today, industry impacts on the organisation of war. Technology has allowed us to develop 'just-in-time warfare'. Here, just like the retail trade, when an item is used up, computers automatically order a new one from supplies, and arrange for its delivery. This way stockpiling becomes redundant, preventing the enemy from establishing where an attack will come from by identifying stockpiles of equipment.
49 A. Rapoport, 'Introduction to On War', in C. Von Clausewitz, *On War*, London: Penguin Classics, 1982, p. 24.
50 Howard, op. cit., p. 96.
51 Ibid., p. 110.
52 For a detailed account of how monarchs used mercenaries and privateers to further their interests beyond their immediate borders, see Thomson, op. cit.

53 In certain instances, the close relationships, which made up these informal networks, were established as far back as when these individuals were at university. For an account of how close these working relationships became, see T. Bower, *The Perfect English Spy*, London: Heinemann, 1995, Chapter 9.

54 The web of informal networks stretched across the channel to France. On at least two occasions, French mercenaries were approached to conduct operations that served British interests. Again, such moves helped further separate the secret world from liberal Britain, see S. Dorril, *MI6: Fifty Years of Special Operations*, London: Fourth Estate, 2000, pp. 685–6 and 736.

55 J. Bloch and P. Fitzgerald, *British Intelligence and Covert Action*, London: Junction Books Ltd, 1983, p. 47.

56 The best example of how close the links were between the commercial security companies and the government was David Stirling's company Watchguard. See A. Hoe, *David Stirling SAS*, London: Warner Books, 1999.

57 Bloch and Fitzgerald, op. cit., p. 50.

58 The force Nasser sent to the Yemen in September 1962 totalled 12,000. By 1965, Egyptian strength was estimated to be 68,000. All these troops were tied up in the Yemen, trying to extract themselves from the mountains. Thus, they were not able in any way to interfere with the British Army's withdrawal from Aden, nor threaten the Saudis to the north. At the same time, the mercenary structure for the Yemen effort totalled approximately 48 ex-servicemen, of whom most were British ex-SAS personnel. See Hoe, op. cit., pp. 365–8.

59 Bloch and Fitzgerald, op. cit., p. 46.

60 Bower, op. cit., p. 219.

61 Ibid., p. 242.

62 Julian Amery had been a member of the Special Operations Executive (SOE) during the Second World War. He had operated in the Middle East and the Adriatic, as well as being parachuted into Albania in 1944. He had also worked alongside McLean leading guerrillas against Germans in Albania in 1944. Both had suffered hardship, enjoyed adventure, and displayed considerable courage in combat. Both played crucial roles in establishing a mercenary force in Yemen to counter the Republican and Egyptian troops. Bower, op. cit., pp. 244–50.

63 Ibid., p. 244.

64 Inward Telegram, To the Secretary of State for the Colonies, PREM 11/3878.

65 Founded in 1736, White's is the oldest club in London. Clubs were, and still are frequently used to conduct this type of business because of the private atmosphere they offer.

66 Bower, op. cit., p. 246.

67 Ibid., p. 249.

68 Hoe, op. cit., p. 358.

69 Ibid., pp. 361–3.

70 Ibid., p. 364.

71 The British government acknowledged that mercenaries might be helping the Royalists, but were acting against government wishes and beyond their control. As noted by Bloch and Fitzgerald, this approach allowed the government to maintain its international reputation while benefiting from the help the mercenaries gave the Royalists, and in the event of something going wrong to be able to deny having anything to do with them. The Foreign Office, Telegram No. 374, PREM 11/4929.

72 Ibid., p. 365.

73 Ibid., p. 366.

74 Prime Minister's Office, Letter on Yemen, PREM 11/4928.

75 The Suez crisis caused a rift in the 'special relationship' between London and Washington when Washington refused to give support to the seizure of the canal by British, French, and Israeli troops after Nasser had nationalised it. The rift between

both sets of policy makers was rapidly repaired in 1957–8 following the recognition by them that the main beneficiary of this rift had probably been the Soviet Union. See D. Sanders, *Losing an Empire, Finding a Role*, London: Macmillan Press Ltd, 1990, pp. 88–96.

76   Hoe, op. cit., p. 357.
77   Ibid., pp. 371–2.
78   Ibid., p. 372.
79   Bloch and Fitzgerald, op. cit., p. 47.
80   Ibid., p. 48.
81   Ibid., p. 48.
82   Hoe, op. cit., p. 404.
83   Ibid., p. 405.
84   Dorril, op. cit., p. 735.
85   The operation was supposed to liberate 140 political prisoners in Tripoli. After their release they were to be left arms while the mercenaries departed. The prison was known as the 'Tripoli Hilton' – hence the press dubbing it the 'Hilton Assignment'. See Hoe, op. cit., pp. 410–11.
86   Dorril, op. cit., p. 735.
87   Ibid., p. 736.
88   Ibid., p. 735.
89   Ibid., p. 735.
90   Ibid., p. 736.
91   H. Kissinger, *Years of Renewal*, London: Weidenfeld & Nicholson, 1999, p. 795.
92   A.-F. Musah and J. Fayemi, 'Africa in Search of Security: Mercenaries and Conflicts an Overview', A.-F. Musah and J. Fayemi (eds), *Mercenaries*, London: Pluto Press, 2000, p. 23.
93   Kissinger, op. cit., p. 812.
94   The '40 Committee' was a committee of the National Security Council, which had historically reviewed covert programmes before they were submitted to the President for approval. Ibid., p. 317.
95   C. Andrew, *For the President's Eyes Only*, London: Harper Collins, 1995, pp. 411–12.
96   Bloch and Fitzgerald, op. cit., p. 194.
97   Callan was dishonourably discharged from the British Army, and jailed for five years, for an armed raid on a Northern Ireland post office, while serving in the Parachute Regiment.
98   A. Mockler, *The New Mercenaries*, London: Corgi, 1986, pp. 227–9.
99   Peter McAleese, who served as a mercenary in Angola, makes this same point in his book. After arriving in the country he explained how all these people had been brought together at short notice, with no allegiance to each other. P. McAleese, *No Mean Soldier*, London: Orion, 1998, p. 88.
100 Interview with Lord Westbury, Chief Executive, Hart GMSSCO Cyprus Ltd, 1 April 2004.
101 K. Connor, *Ghost Force: The Secret History of the SAS*, London, Orion, 1999, p. 306.
102 Ibid.
103 T. Geraghty, *Who Dares Wins*, London, Time Warner, 2002, p. 424.
104 Interview with Lord Westbury, 1 April 2004.
105 Kroll was founded in 1972 by Jules B. Kroll initially to offer investigative and security services to corporate clients. Kroll, http://www.krollworldwide.com/ (accessed 1 November 2004). Control Risks Group was founded in 1975 by David Walker and three fellow ex-British SAS men. Walker, along with Jim Johnson formed Keenie Meenie Services (KMS) in the early 1980s after leaving Control Risks Group. Saladin Security was founded as a subsidiary of KMS to operate domestically while

KMS concentrated on international contracts. Saladin International began to operate internationally after KMS folded in the early 1990s. DSL was founded in 1981 by Alastair Morrison, a former SAS officer.

106 M. Kaldor, *New & Old Wars*, Cambridge: Polity Press, 2002, p. 1.
107 Ibid., p. 6.
108 Ibid., p. 7.
109 D. Keen, 'The Economic Function of Violence in Civil Wars', *Adelphi Paper 320*, London: International Institute for Strategic Studies, 1998, p. 11.
110 Kaldor, op. cit., p. 5.
111 M. Duffield, *Global Governance and the New Wars*, London: Zed Books, 2001, p. 121.
112 Ibid., p. 121.
113 Ibid., p. 121.
114 Ibid., p. 15.
115 C. Tilly, *Coercion, Capital, and European States: AD 990–1992*, Oxford: Blackwell, 1990, pp. 69–70.
116 M. Duffield, op. cit., p. 63.
117 A.-F. Musah, 'A Country Under Siege: State Decay and Corporate Military Intervention in Sierra Leone', in Musah and Fayemi (eds), op. cit., p. 88.
118 D. Shearer, 'Private Armies and Military Intervention', *Adelphi Paper 316*, London: International Institute for Strategic Studies, 1998, p. 49.
119 T. Catán and S. Fidler, 'The Military can't Provide Security. It had to be Outsourced to the Private Sector and that was our Opportunity', in *The Financial Times*, 29 September 2003.
120 Ibid.

3 THE ROLE OF PMCs AFTER THE COLD WAR

1 Foreign and Commonwealth Office, *Private Military Companies: Options for Regulation*, London: The Stationery Office, 2002, pp. 4–5.
2 L. Martin and J. Garnett, *British Foreign Policy: Challenges and Choices for the 21st Century*, London: The Royal Institute of International Affairs, 1997, pp. 83–5.
3 A very good example of the West's general lack of interest in Africa, outside of normal trade relations, can be identified in its response to the civil war in Rwanda. L. Melvern, *A People Betrayed*, London: Zed Books, 2000, p. 233.
4 Ibid., p. 233.
5 N. Wheeler, *Saving Strangers: Humanitarian Intervention in International Society*, Oxford: Oxford University Press, 2000, p. 221.
6 For an account of how sovereignty can be used to legitimate or discourage intervention, see S. Krasner, *Sovereignty: Organised Hypocrisy*, Princeton, NJ: Princeton University Press, 1999.
7 M. Duffield, *Global Governance and the New Wars*, London: Zed Books, 2001, p. 65.
8 J. Hirsch, *Sierra Leone: Diamonds and the Struggle for Democracy*, London: Lynne Rienner, 2001, p. 114.
9 D. Shearer, 'Private Armies and Military Intervention', *Adelphi Paper 316*, London: International Institute for Strategic Studies, 1998, p. 62.
10 W. Reno, *Warlord Politics and African States*, London: Lynne Rienner, 1998.
11 P. Singer, *Corporate Warriors: The Rise of the Privatised Military Industry*, London: Cornell University Press, 2003, p. 102.
12 Ibid., p. 102.
13 K. Pech, 'Executive Outcomes: A Corporate Conquest', in J. Cilliers and P. Mason, (eds), *Peace, Profit or Plunder*, South Africa: Institute for Security Studies, 1999, p. 85.

14 Shearer, op. cit., p. 46.
15 Pech, op. cit., p. 85.
16 Shearer, op. cit., p. 46.
17 Ibid., p. 46.
18 The international community did give humanitarian assistance throughout the Civil War, while a number of countries did supply weapons to the Angolan government, though such supplies probably fuelled the fighting. Human Rights Watch, *Angola: Arms Trade and Violations of the Laws of War Since the 1992 Elections*, London: Human Rights Watch, 1994.
19 Shearer, op. cit., p. 46.
20 Ibid., p. 51.
21 Hirsch, op. cit., p. 114.
22 Economic and Social Council, Report E/CN.4/1998/31, 1998.
23 K. O'Brien, 'Private Military Companies and African Security 1990–1998', in A-F. Musah and K. Fayemi, *Mercenaries*, London: Pluto Press, 2000, pp. 44–5.
24 T. Spicer, 'Interview', *Cambridge Review of International Relations*, Cambridge: 1999, Vol. XIII, No. 1, p. 166.
25 Duffield, op. cit., p. 66.
26 Shearer, op. cit., p. 24.
27 Ibid., p. 36.
28 Duffield, op. cit., p. 66.
29 S. Dorril, *MI6: Fifty Years of Special Operations*, London: Fourth Estate, 2000, p. 61.
30 Shearer, op. cit., p. 36.
31 Dorril, op. cit., pp. 736–7.
32 Some major differences between these two groups include the latter's provision of a legally binding service contract, a code of practice that ensures risks taken by the company are appropriately covered. In respect to the last item, insurance companies are able to demand that a company employee behaves in a particular way for that person to be covered, and that acts as a useful additional compliance discipline, as employees usually want to ensure that they are personally covered. Whether this type of detail would apply to a mercenary contract is questionable, especially in relation to the question of insurance. Interview with Nigel Woof, 30 November 2001.
       For an idea of what was involved in a contract between an employer and mercenary fighting in the Congo in the 1960s see A. Mockler, *The Mercenaries*, New York, NY: Macmillan, 1986, p. 465.
33 Pech, op. cit., p. 83.
34 The registered number of EO is 95/02016/07, indicating that it was only registered in 1995. It is possible that the company was only registered in England in 1995, while being registered much earlier in South Africa. S. Cleary, 'Angola: A Case Study of Private Military Involvement', in J. Cilliers and P. Mason (eds), *Peace, Profit or Plunder*, South Africa: Institute for Security Studies, 1999, p. 170.
35 Pech, op. cit., p. 84.
36 Ibid., p. 84.
37 For a list of affiliated companies see Pech, op. cit., pp. 87–8.
38 For an account of the ethos behind Sandline International see Spicer, op. cit., pp. 151–2.
39 Spicer, op. cit., p. 165.
40 Spicer, op. cit., p. 165.
41 P. McAleese, *No Mean Soldier*, London: Orion, 1998, p. 280.
42 J. Cilliers and I. Douglas, 'The Military as Business: Military Professional Resources, Incorporated', in Cilliers and Mason (eds), op. cit., pp. 112–13.
43 Ibid., p. 113.
44 Ibid., p. 112.

45 Ibid., p. 114.
46 E. Soyster (Lt Gen. Retd), Interview, Washington, DC, Center for Defense Information. Online. Available at http://www.cdi.org/adm/1113/Shoyster.html (accessed 7 March 2002).
47 Ibid.
48 Cilliers and Douglas, op. cit., Chapter 6.
49 Even Group 4 Falck have been charged with being a mercenary outfit after one of its officers of South African origin was taken hostage by the RUF. No suggestion was made that the person took part in combat or support operations, only that he had been taken hostage. A-F. Musah, 'A Country Under Siege', Musah and Fayemi, op. cit., p. 88.
50 In the passage Pech refers only to EO. Even so, the general line her argument takes suggests all companies similar to EO are able to mutate. See Pech, op. cit., p. 84.
51 Shearer, op. cit., p. 36.
52 Sir Thomas Legg and Sir Robin Ibbs, *Report of the Sierra Leone Arms Investigation*, London: The Stationery Office, 1998.
53 Ibid., p. 4.
54 Foreign Affairs Committee, *Second Report*, London: The Stationery Office, 1999, p. xiv.
55 Singer, op. cit., p. 109.

4 TENSIONS EXPOSED IN THE 'ARMS FOR AFRICA' AFFAIR

1 An indication of the feeling of abhorrence towards mercenaries and the need to regulate PMCs is given in the debates on the subject, and can be gauged from statements by Robin Cook, Tony Lloyd, Donald Anderson, Paddy Ashdown, Menzies Campbell, and Andrew Mackinlay. Paddy Ashdown (13 May 1998) *Hansard*, pt 18, column 370. Menzies Campbell (18 May 1998) *Hansard*, pt 10/11, column 614/617. Robin Cook (27 July 1998) *Hansard*, pt 7, column 27. Robin Cook (27 July 1998) *Hansard*, pt 7, column 27. Andrew Mackinlay (1 December 1998) *Hansard*, pt 1, column 670. Tony Lloyd (1 December 1998) *Hansard*, pt 1, column 670. Robin Cook (2 March 1999) *Hansard*, pt 9, column 895. Menzies Campbell (2 March 1999) *Hansard*, pt 12, column 903. Tony Lloyd (2 March 1999) *Hansard*, pt 20, column 932. What is common to a number of these statements is a perception of a need to regulate PMCs to control their activities is already recognised as the way ahead for these companies.
2 The switch from using the term mercenary in the House of Commons debates to PMC in the Green Paper, which still refers to mercenaries, is particularly interesting in that it appears a conscious effort is being made early on to separate these two groups by the government.
3 An indication of government attitude towards PMCs is included in the Foreword to the Green Paper by the Secretary of State for the FCO for the government's attitude towards PMCs. Foreign and Commonwealth Office, *Private Military Companies: Options for Regulation*, London: The Stationery Office, 2002.
4 See fn. 1.
5 ECOMOG is the military arm of Economic Community of West African States (ECOWAS).
6 The chronology of selected events in Sierra Leone given here is taken from J. Hirsch, *Sierra Leone: Diamonds and the Struggle for Democracy*, London: Lynne Rienner, 2001, p. 113, and Foreign Affairs Committee, 'Second Report', *Sierra Leone*, London: The Stationery Office, 1999, p. 5.
7 Hirsch, op. cit., p. 66.
8 Interview with Michael Grunberg, Official Spokesman for Sandline International, 24 June 2002.

9 Sir Thomas Legg and Sir Robin Ibbs, *Report of the Sierra Leone Arms Investigation*, London: The Stationery Office, 1998, paragraph 4.6, p. 24.

10 Ibid., paragraph 3.5, p. 14.

11 Foreign Affairs Committee, op. cit., paragraph 11, p. xiii.

12 Sir Thomas Legg and Sir Robin Ibbs, op. cit., paragraph 3.4, pp. 13–14.

13 United Nations Security Council Resolution No. 1132, 1997, paragraph 3.

14 Article 41 of the Charter allows the Security Council to decide on what measures not involving the use of armed force to employ, while Article 42 does allow the use of armed force to maintain or restore international peace and security. A. Roberts and B. Kingsbury (eds), *United Nations, Divided World*, Oxford: Clarendon Press, 1995, p. 511.

15 United Nations Security Council Resolution No. 1132, 1997, paragraph 6.

16 Sir Thomas Legg and Sir Robin Ibbs, op. cit., paragraph 3.15, p. 17.

17 Foreign Affairs Committee, op. cit., paragraph 13, p. ix.

18 United Nations Security Council Resolution No. 1132, 1997, paragraph 7.

19 See fn. 16 for text of paragraph 6.

20 Foreign Affairs Committee, op. cit., paragraph 14, p. x.

21 Ibid., p. x, paragraph 16.

22 *Hansard*, Volume 308, Sierra Leone, 12 March 1998.

23 Sir Thomas Legg and Sir Robin Ibbs, op. cit., paragraph 3.9, p. 15.

24 Ibid., paragraph 3.19, p. 18.

25 Ibid., paragraph 3.24, p. 20.

26 Foreign Affairs Committee, op. cit., paragraph 21, p. xii.

27 Ibid., paragraph 21, p. xii.

28 Hirsch, op. cit., pp. 65–73.

29 Interview with Michael Grunberg, 24 June 2002.

30 Sir Thomas Legg and Sir Robin Ibbs, op. cit., paragraph 4.2, p. 23.

31 Sandline International, http://www.sandline.com (accessed 16 December 2002).

32 Hirsch, op. cit., p. 66.

33 Ibid., p. 66.

34 E. O'Loughlin, 'Sandline Scandal Arms Shipment Reaches Forces', *Independent*, 22 May 2000, p. 4.

35 Ibid., p. 4.

36 Sir Thomas Legg and Sir Robin Ibbs, op. cit., paragraph 4.10, p. 25.

37 M. Colvin and N. Rufford, 'Our Man in Sierra Leone is the People's Hero', *The Times*, 17 May 1998. In the following article, Saxena stated, '... the Grantee [Saxena] agrees to give economic and other assistance, to the value of US$10m to the Grantor [Kabbah] for ... restoration of the Constitution'.

38 Sir Thomas Legg and Sir Robin Ibbs, op. cit., paragraph 4.15, p. 26.

39 Interview with Michael Grunberg, 24 June 2002.

40 Graduate Institute of International Studies, Geneva, *Small Arms Survey 2001: Profiling the Problem*, Oxford: Oxford University Press, 2001, p. 104.

41 Interview with Michael Grunberg, 24 June 2002.

42 Sir Thomas Legg and Sir Robin Ibbs, op. cit., paragraph 4.18, p. 27.

43 Graduate Institute of International Studies, Geneva, op. cit., p. 112.

44 Ibid., p. 171.

45 Sir Thomas Legg and Sir Robin Ibbs, op. cit., paragraph 2.1, p. 7.

46 The Berwin Letter, http://www.fco.gov.uk (accessed 29 October 2004).

47 Ibid., pp. 7–9.

48 Ibid., p. 11.

49 M. Kaldor, *New & Old Wars*, Cambridge: Polity Press, 2001, p. 2.

50 Sir Thomas Legg and Sir Robin Ibbs, op. cit., paragraph 6.17, p. 52, and paragraph 6.25, p. 54.

51 For an account of what Spicer claimed to have told FCO officials about the arms transfer, see T. Spicer, *An Unorthodox Soldier*, Edinburgh: Mainstream Publishing, 1999, Chapter 16.

52 Sir Thomas Legg and Sir Robin Ibbs, op. cit., paragraph 6.21, p. 53.

53 Interview with Michael Grunberg, 24 June 2002.

54 The FCO would have known about the company's involvement in Sierra Leone as early as 10 December 1997 when Buckingham, a director of Sandline International, and Bowen, Sandline International's representative in Conakry, visited the FCO. See Spicer, op. cit., p 212.

55 Sir Thomas Legg and Sir Robin Ibbs, op. cit., paragraph 4.14, p. 26.

56 Ibid., paragraph 1.1, p. 1.

57 S. Makki, S. Meek, A-F. Musah, M. Crowley and D. Lilly, *Private Military Companies and the Proliferation of Small Arms: Regulating the Actors*, London: International Alert, 2000, p. 4.

58 Donald Anderson was the Chairman of the Foreign Affairs Committee, and has spoken extensively on the 'arms to Africa' affair.

59 Anderson, *Hansard*, pt 12, column 622, 18 May 1998.

60 The full range of the company's activities was discussed during the author's interview with Michael Grunberg, 24 June 2002.

61 Sir Thomas Legg and Sir Robin Ibbs, op. cit., paragraph 1.5, p. 4.

62 Ibid., paragraph 1.1, p. 3.

63 Ibid., paragraph 10.16, p. 107.

64 Ibid., paragraph 6.22, pp. 53–4.

65 Interview with Michael Grunberg, 24 June 2002.

66 Ibid.

67 *The Times* (1998), *Sandline is given support by the UN*. Online. Available at http://www.times-archive.co.uk/news/pages/tim/98/05/23/timfgnfgn01001.html (accessed 29 October 2004).

68 Interview with Michael Grunberg, 24 June 2002.

69 This is very possible especially as Spicer himself points to this very problem, explaining that 'it [was] clear the UK's Order in Council was flawed and did not accurately reflect the words or the intentions of the original UN Security Council Resolution on which it was based and from which it draws its authority'. Spicer, op. cit., p. 204.

70 Foreign Affairs Committee, op. cit., paragraph 22, p. xii.

71 Sir Thomas Legg and Sir Robin Ibbs, op. cit., paragraph 3.28, p. 21.

72 Spicer, op. cit., p. 205.

73 Foreign Affairs Committee, op. cit., paragraph 23, p. xiii.

74 Sir Thomas Legg and Sir Robin Ibbs, op. cit., paragraph 3.28, p. 21.

75 Foreign Affairs Committee, op. cit., paragraph 24, p. xiii.

76 Ibid., paragraph 24, p. xiii.

77 Ibid., paragraph 27, p. xv.

78 Sir Thomas Legg and Sir Robin Ibbs, op. cit., Annex A, p. 120.

79 Two debates were held during opposition days on Sierra Leone, the first on 18 May 1998, the second on 2 March 1999. A further debate was held by the Legg enquiry on 27 July 1998.

80 Interview with Michael Grunberg, 24 June 2002.

81 Interview with Michael Grunberg, 24 June 2002.

82 Foreign and Commonwealth Office, op. cit., p. 5.

83 For a brief account of the laws affecting mercenaries, see C. Beyani and D. Lilly, *Regulating Private Military Companies: Options for the UK Government*, London: International Alert, 2001, pp. 21–33.

84 The Green Paper originated after receiving a request from the Foreign Affairs Committee of the House of Commons. The remit for the paper was to look into the

whole issue of PMCs, to see if they could contribute to British foreign policy in future.

85 Foreign and Commonwealth Office, op. cit., p. 27.

86 Both these countries have passed legislation that regulates the actions of all types of commercial security companies. The 1998 South African Regulation of Foreign Military Assistance Act is the most far-reaching national legislation dealing with mercenaries and private PMCs. In the US, control over arms brokering and the export of military services are dealt with in a similar way, through the International Traffic in Arms Regulations (ITAR), overseen by the Department of State's Office of Defence Trade Control.

87 The report was concerned only with matters that related directly to the questions about how much British officials and ministers knew of the allegations about the supply of arms to Sierra Leone, and how far they gave encouragement and approval, and with what authority. Wider questions to do with the supplying of military services by Sandline that could have suggested a close association with classic mercenaries were not discussed. See Sir Thomas Legg and Sir Robin Ibbs, op. cit., p. 8, paragraph 2.3.

88 Interview with Michael Grunberg, 24 June 2002.

89 Ibid., paragraph 4.17, p. 27.

90 Foreign and Commonwealth Office, op. cit., paragraph 59, p. 19.

91 Ibid., pp. 4–5.

5 THE PRIVATISATION OF WARFARE AND THE EMERGENCE OF PMCs

1 F. Rohatyn and A. Stranger, 'The Profit Motive Goes to War', in *The Financial Times*, 17 November 2004.

2 The Center for Public Integrity, http://www.publicintegrity.org (accessed 23 December 2004).

3 J. Tamayo, 'Pentagon contractors Playing Vital Military Support Roles'. Online. Available at http://www.miami.com/mld/miamiherald/news/nation/5335374.htm (accessed 24 December 2004).

4 BBC News. Online. Available at http://www.news.bbc.co.uk/2/hi/uk_news/4098917.stm (accessed 24 December 2004).

5 For an analysis of the directive, see A. Roberts, 'The Crisis in UN Peacekeeping', in C. Crocker and F. Hampson (eds), *Managing Global Chaos*, Washington, DC: Institute of Peace, 1996.

6 H. Howe, *Ambiguous Order: Military Forces in African States*, London: Lynne Rienner Publishers, 2001, p. 111.

7 M. Ignatieff, 'The Gods of War', *New York Review of Books,* 1997, p. 13.

8 KP normally refers to kitchen porter duties.

9 M. Fineman, 'Privatised Army in Harm's Way', in *Los Angeles Times*, 24 January 2003.

10 S. Leslie, *The Cold War and American Science*, New York, NY: Columbia University Press, 1993, p. 1.

11 Ibid., p. 2.

12 C. Moskos, 'Towards a Postmodern Military: The United States', in C. Moskos, *et al.*, *The Post Modern Military: Armed Forces after the Cold War*, Oxford: Oxford University Press, 2000, p. 21.

13 All the information on the following companies has been taken from their official websites. Booz Allen Hamilton, http://www.boozallen.com (accessed 4 December 2002).

14 Vinnell Corporation, http://www.vinnell.com (accessed 23 March 2005).

15 DynCorp, http://www.dyncorp.com (accessed 23 March 2005).

16 TRW Corporation, http://www.trw.com (accessed 23 March 2005).

17 SAIC, http://www.saic.com (accessed 23 March 2005).
18 J. Kurlantzick, 'Outsourcing the Dirty Work: The Military and its Reliance on Hired Guns', in *The New Republic*, 24 April 2003.
19 Interestingly, the US Army Intelligence and Security Command put together a team of five civilian contractors, who flew and maintained UAVs in Central America in 1984. Interview with Ed Soyster, Vice President MPRI, 10 September 2002.
20 Editorial, 'DOD Considers Using Civilians to Pilot UAV in Military Operations', in *Inside the Air Force*, 1 November 2002.
21 PR News Association Inc., 'AirScan Order 30 Adam Aircraft A500s for Airborne Surveillance', 20 December 2002.
22 R. Little, 'American Civilians Go Off To War', in *Baltimore Sun*, 26 May 2002.
23 G. Roseanne, 'L-3 tapped to aid Army intel command'. Online. Available at http://www.washingtontechnology.com/news/1_1/daily_news/25199-1.html (accessed 4 January 2005).
24 CICA's job descriptions are detailed in two task orders and can be viewed online at http://www.publicintegrity.org/wow/report.aspx?aid=361 (accessed 4 January 2005).
25 Ibid.
26 Online. Available at http://www.globalsecurity.org/military/library/news/2002/07/mil-0020725-361ee43c.htm (accessed 28 January 2003).
27 Interview with Ed Soyster, Vice President, MPRI, September 2002.
28 Interview with Ed Soyster, 10 September 2002.
29 In making MPRI an exception, McGhie is probably thinking of MPRI's role in training overseas armies, in particular the Croat and Bosnian armies, see D. Shearer, 'Private Armies and Military Intervention' in, *Adelphi Paper 316*, International Institute for Strategic Studies, 1998.
30 S. McGhie, 'Private Military Companies: Soldiers, Inc', in *Jane's Weekly Defence*, London: Thanet Press Ltd, Vol. 37, No. 21, 2002.
31 Howe, op. cit., p. 220.
32 All contracts undertaken by MPRI on behalf of the US government come under the direct control of a government official who is responsible, and thus accountable, for ensuring MPRI teaches only what is in the contract. All MPRI's contracts are also subject to 'Freedom of Information Act'. Interview with Ed Soyster, 10 September 2002.
33 C. Lee, 'Army Weighs Privatisation Close to 214,000 Jobs One in Six Workers Could be Affected', in *The Washington Post*, 3 November 2002.
34 Shearer, op. cit., pp. 35–6.
35 T. Spicer, 'Interview', in *Cambridge Review of International Affairs*, Vol. XIII, No. 1, 1999, p. 165.
36 Shearer, op. cit., p. 36.
37 McGhie, op. cit., pp. 24–5.
38 Shearer, op. cit., p. 36.
39 J. Burns and A. Parker, 'Ex-SAS soldiers hope for new lease of life', in *The Financial Times*, 14 February 2002.
40 Significant numbers of former Guards Officers work for, or own, UK PMCs.
41 Interview with Kevin James, 19 January 2002.
42 In an interview with an official from ArmorGroup on 30 November 2001, the official made it quite clear that they normally only employ former members of 33 Regiment Royal Engineers on EOD operations.
43 Interview with Christopher Beese, Chief Administrative Officer, ArmorGroup, 16 November 2004.
44 Interview with Kevin James, 19 January 2002.
45 S. Fidler, 'The Military can't Provide Security. It had to be Outsourced to the Private Sector and that was Our Opportunity', in *The Financial Times*, 29 September 2003.

46 J. Burns and J. Drummond, 'Controversial Ex-British Army Officer Given Key Iraq Post', in *The Financial Times*, 19 June 2004.

47 Interview with Robert Ashington-Pickett, Director, Control Risks Group, 16 November 2004.

48 Agence France Presse, 'Britain Banks on Private Security Firms in Iraq as Civilian Gunned Down', 28 March 2004.

49 Babcock International, http://www.babcock.co.uk/ (accessed 6 March 2005).

50 Ministry of Defence, http://www.mod.uk/business/ppp/ppp_defence.htm (accessed 6 March 2005).

51 W. Pincus, 'More Private Forces Eyed for Iraq: Green Zone Contractor Would Free US Troops for Other Duties', in *Washington Post*, 18 March 2004.

52 M. Fitzgerald, 'US Contract to British Firm Sparks Irish American Protest', in *Washington Post*, 9 August 2004.

53 The Army invalidated four other competing proposals, one of which was from the much larger and more experienced firm, DynCorp International LLC. The company also had its protest, that its proposal was unreasonably evaluated and improperly excluded from consideration for award appeal, later turned down.

54 B. Brady, 'Hundreds of Army Officers Resign to Cash in on Iraqi Security Boom', in *Scotland on Sunday*, 25 April 2004.

55 Editorial, 'Soldiers Leave for Iraqi Jobs', in *Fiji Times*, 23 December 2003.

56 D. Wood, 'Some of Army's Civilian Contractors Are No-Shows in Iraq'. Online. Available at http://www.newhousenews.com/archive/wood080103.html (accessed 11 January 2005).

57 Ibid.

58 E. Schrader, 'US Companies Hired to train Foreign Armies', in *Los Angeles Times*, 14 April 2002.

59 S. Lunday, 'Firms Join Security Drive', in *Charlotte Observer*, 13 February 2002.

60 Available at http://www.smh.com.au/articles/2003/11/20/1069027266818.html (accessed 12 December 2005).

61 Schrader, op. cit.

62 P. Singer, *Corporate Warriors: The Rise of the Privatised Military Industry*, London: Cornell University Press, p. 232.

6 THE ROLE OF PMCS IN A CHANGING GLOBAL ENVIRONMENT

1 M. Kaldor, *New & Old Wars*, Cambridge: Polity Press, 2002, p. 92.

2 K. O'Brien, 'Private Military Companies and African Security', in A-F. Musah and J. Fayemi (eds), *Mercenaries*, London: Pluto Press, 2000, pp. 58–9.

3 M. Duffield, *Global Governance and the New Wars*, London: Zed Books, 2001, p. 12.

4 Ibid.

5 Ibid., p. 139.

6 W. Reno, 'Shadow States and the Political Economy of Civil Wars', in M. Berdal and D. Malone (eds), *Greed and Grievance: Economic Agendas in Civil Wars*, London: Lynne Rienner, 2000, p. 45.

7 Duffield, op. cit., p. 166.

8 Ibid., p. 167.

9 W. Reno, *Warlord Politics and African States*, London: Lynne Rienner Publishers, 1999, p. 97.

10 Duffield, op. cit., p. 145.

11 Global Witness, 2002, *Press release*, London: Global Witness. Online. Avaliable at http://www.globalwitness.org/press_releases/pressreleases.php?type=liberia (accessed 19 June 2002).

12 R. Abrahamsen, *Disciplining Democracy*, London: Zed Books, 2000.

13 Reno, op. cit., p. 3.

14 Ibid., p. 3.

15 Duffield, op. cit., p. 175.

16 Ibid., p. 175.

17 ECOMOG is a military force formed by member states of the ECOWAS.

18 For a detailed account of Taylor's involvement with foreign firms see, Reno, op. cit., chapter 3.

19 Duffield, op. cit., p. 176.

20 For a detailed account of Liberia's civil war, see Reno, op. cit., chapter 3.

21 The West Side Boys (men, women, and children) were a pro-government militia commanded by the self-styled 'Brigadier' Foday Kallay. The group was a thorn in the side of a government trying to establish peace and security in the country. They frequently took local people prisoners, including women whom they forced into prostitution, as well as plundering local businesses or any other source of wealth. British Special Forces and the Parachute Regiment, sent in to rescue a detachment from the Royal Irish Regiment held hostage by the group, eventually dispersed, killed or captured most of the group members. Foday Kallay was captured during the rescue mission. S. Kiley, 'West Side Boys regroup under threat of death', in *The Times*, 14 September 2000.

22 For an account of the role played by the Kamajohs in the Sierra Leone civil war see, J. Hirsch, *Sierra Leone: Diamonds and the Struggle for Democracy*, London: Lynne Rienner Publishers, 2001, p. 52.

23 Chief Hinga Norman subsequently became Kabbah's deputy defence minister; see Hirsch, op. cit., p. 39.

24 Ibid., p. 52.

25 For an account of the second Congo war, see F. Reyntjens, 'The Second Congo War: More Than a Remake', in *African Affairs*, Oxford: Oxford University Press, Vol. 98, No. 391, 1999.

26 Zimbabwe appears to have been the only intervening country whose intentions for doing so were economic. At the end of the 1996–7 war, they were concerned about repayment in the event of Kabila being overthrown. A second economic motive lay in Zimbabwean business interests in the Congolese market, especially the mining sector. Reyntjens, op. cit., p. 248.

27 As is stated in one of International Alert's numerous papers published on private security, countries seeking military assistance are more likely to turn to other armies in Africa than to mercenaries. See T. Vaux *et al.*, *Humanitarian Action and Private Security Companies*, London: International Alert, 2002, p. 11.

28 During the Civil War in Zaire (now the Democratic Republic of Congo) Mobutu considered hiring PMCs to rescue his Presidency. P. Thornycroft, 'Mobutu couldn't afford SA Mercenaries', in *Electronic Mail & Guardian*, 18 July 1997.

29 K. O'Brien, 'Private Military Companies and African Security', in A-F. Musah and J. Fayemi (eds), *Mercenaries*, London: Pluto Press, 2000.

30 For an account of private military forces that took part in the civil war in Zaire from 1996–7, see O'Brien, op. cit.

31 Thornycroft, op. cit.

32 D. Keen, 'The Economic Function of Violence in Civil Wars', in *Adelphi Paper 320*, London: International Institute for Strategic Studies, 1998.

33 Kaldor, op. cit., pp. 6–9.

34 M. V. Creveld, 'The Transformation of War Revisited', in R. J. Bunker (ed.), *Non-State Threats and Future Wars*, London: Frank Cass, 2003, p. 8.

35 For example, Kaldor understands new wars as representing a mixture of war, crime, and human rights violations. Kaldor, op. cit., p. 11.
36 Duffield, op. cit., p. 188.
37 Ibid., p. 13.
38 Ibid., p. 194.
39 Reno, op. cit.
40 Duffield, op. cit., p. 194.
41 A big problem for weak states is the lack of a countrywide infrastructure and other important medical and social facilities. The mineral extraction companies on the other hand are well equipped to build the type of facilities, including roads, runways, schools, and clinics that local people need. All this is not to suggest that we should allow mineral extraction companies a free hand because of what services they can provide a community. However, if the socio–political conditions are suitable, we should not ignore what these companies can do for local communities. Interview with Brian Spratley, Consultant Mining Engineer, Hythe, UK, 20 April 2001.
42 Reno, op. cit.
43 Duffield, op. cit., p. 11.
44 A. Hurrell, 'Explaining the Resurgence of Regionalism in World Politics', in *Review of International Studies*, Vol. 21, No. 4, 1995.
45 I. Clark, *Globalisation and Fragmentation: International Relations in the Twentieth Century*, Oxford: Oxford University Press, 1997, pp. 1–9.
46 Duffield, op. cit., pp. 14–15.
47 Ibid., p. 12.
48 A. Vines, 'Mercenaries Human Rights and Legality', in A-F. Musah and J. Fayemi, op. cit., 2000, pp. 186–7.
49 Vaux, *et al.*, op. cit.
50 R. Williams, 'Building Stability in Africa: Challenges for the New Millennium', in *Monograph 46*, South Africa: Institute for Security Studies, 2000, p. 2.
51 Ibid., p. 2.
52 Ibid., p. 2.
53 D. Hendrickson, 'A Review of Security Sector Reform', *Working Paper 1*, Centre for Defence Studies, University of London, 1999, p. 9.
54 Williams, op. cit., p. 2.
55 Interview with Robert Ashington-Pickett, London, 16 November 2004.
56 MPRI, http://www.mpri.com/site/capabilities.html (accessed 14 December 2004).
57 Interview with Claire Howard, Atos Consulting, 8 December 2004.
58 Ibid.
59 Interview with Robert Ashington-Pickett, London, 16 November 2004.
60 Interview with Claire Howard, 8 December 2004.
61 Transcript of interview between Eben Barlow and Jim Hooper, 26 January 1996.
62 D. Shearer, 'Private Armies and Military Intervention', in *Adelphi Paper 316*, London: International Institute for Strategic Studies, 1998, p. 68.
63 Duffield, op. cit.
64 Kaldor, op. cit., p. 95, also see M. Bilton, 'Death and Dollars', in *The Sunday Times Magazine*, 2 July 2000, pp. 32–47.
65 I. Douglas, 'Fighting for Diamonds: Private Military Companies in Sierra Leone', in J. Cilliers and P. Mason (eds), *Peace, Profit, or Plunder*, South Africa: Institute for Security Studies, 1999.
66 A-F. Musah and J. Fayemi, 'Africa in Search of Security: Mercenaries and Conflicts: An Overview', in Musah and Fayemi, op. cit., p. 23.
67 Vaux, Seiple, Nakano and Van Brebant, op. cit., p. 16.
68 ArmorGroup, Brochure, London, 2001.
69 It is normal for PMCs to set up a PSC if their contract involves passive security. For example, EO set up Lifeguard, a PSC, to guard the mines in Sierra Leone. Each

company had separate and clearly defined responsibilities even though for the same client. However, because EO used the same pool of retired soldiers to set up the PSC, personnel were transferred between companies.

70 D. Lilly, 'The Privatisation of Security and Peace Building: a Framework for Action', in *International Alert*, London: International Alert, 2000.

71 These skills range from peace enforcement, peacekeeping, mine clearance, support for NGOs, convoy escorts, guarding food distribution points, water purification, and construction work.

72 Transcript of interview between Eeben Barlow and Jim Hooper, 26 January 1996.

73 Ibid.

74 Ibid.

75 International Workshop Summary Report, *The Politicisation of Humanitarian Action and Staff Security: The Use of Private Security Companies by Humanitarian Agencies*, London, International Alert, 2001, p. 4.

76 Ibid., p. 3.

77 For a detailed debate on the consequences and implications for NGOs employing private security companies see, Vaux, *et al.*, op. cit., pp. 16–18.

78 Cilliers and Mason (eds), op. cit.

79 Reno, op. cit.

80 MPRI, at http://www.mpri.com/channels/mission.html (accessed 3 March 2003).

7 REGULATION AND CONTROL OF PRIVATE MILITARY COMPANIES

1 Adam Roberts and Richard Guelff, *Documents of the Laws of War*, Oxford: Oxford University Press, 2000, p. 447.

2 Netherlands Yearbook of International Law, Volume XXII, 1991, Martinus Nijhoff Publishers.

3 J. Lewis, 'Did this ex-MI6 man tip off Straw about Africa coup', in *The Mail on Sunday*, 21 November 2004.

4 For a detailed discussion on the issues surrounding the need to regulate PMCs, see Foreign Affairs Committee, 'Private Military Companies', London: The Stationery Office, 2002; Foreign and Commonwealth Office, 'Private Military Companies: Options for Regulation', *The Green Paper*, London: The Stationery Office, 2002; and C. Beyani and D. Lilly, *Regulating Private Military Companies: Options for the UK Government*, London: International Alert, 2001.

5 For a detailed account of the Foreign Enlistments Act 1870 see Lord Diplock, Sir Derek Walker Smith and Sir Geoffrey de Freitas, *The Diplock Report*, Command 6569 (XXI), London: The Stationery Office, 1976.

6 Foreign and Commonwealth Office 'Private Military Companies: Options for Regulation', *Green Paper*, London: The Stationery Office, 2002, p. 20.

7 Ibid.

8 Ibid.

9 P. Singer, 'War Profit and the Vacuum of Law: Private Military Firms and International Law', *Columbia Journal of Transnational Law Association*, Vol. 42, No. 2, p. 532.

10 Ibid.

11 22 CFR Parts 120–30, International Traffic in Arms Regulation, Section 120.1., US: Department of State.

12 Ibid., Section 126.1.

13 D. Avant, 'Privatising Military Training', in *Foreign Policy in Focus*, May 2000.

14 P. Singer, op. cit., p. 538.

15 Ibid., p. 539.

16 The Prohibition of Mercenary Activity and Prohibition and Regulation of Certain Activities in an Area of Armed Conflict will replace the 1999 Foreign Military

Assistance Act which the government considered too weak to secure enough convictions against mercenaries.

17 Republic of South Africa Government Gazette, No. 15 of 1998: Regulation of Foreign Military Assistance Act 1998.

18 Ibid.

19 M. Malan and J. Cillers, 'Mercenaries and Mischief: The Regulation of Foreign Military Assistance Bill', *Occasional Paper No. 25*, Institute for Security Studies, South Africa, 1997.

20 A. Kotze, 'South Africans streamlining into Iraq'. Online. Available at http://www.news24.com/News24/South_Africa/News/0,6119,2-7-1442_1484057,00.html (accessed 25 November 2004).

21 Edited by T. Steyn, 'War Hero: I was a Mercenary', Online. Available at http://www.sandline.com/hotlinks/News24_War-hero.html (accessed 25 November 2004).

22 Foreign and Commonwealth Office, op. cit., pp. 4–5.

23 Ibid., pp. 4–5.

24 C. Adams, 'Straw to Back Controls over British Mercenaries', in *The Financial Times*, 2 August 2002.

25 Menzies Campbell, Response to Green Paper, unpublished documents, August 2002.

26 Foreign and Commonwealth Office, op. cit., p. 22.

27 *United Kingdom Strategic Export Controls Annual Report 2003*, London: The Stationery Office.

28 Lord Diplock, Sir Derek Walker Smith and Sir Geoffrey de Freitas, *Report of the Committee of Privy Counsellors appointed to inquire into the recruitment of mercenaries*, Cmnd. 6569 (XXI), London: The Stationery Office, 1976, p. 3.

29 Ibid., p. 3.

30 Ibid., p. 4.

31 Foreign and Commonwealth Office, op. cit., p. 23.

32 The relevance of this argument depends on an *a priori* value judgement about the nature of the problem created by the recruitment of mercenaries. Only if one takes the view that mercenary activity is a lesser order issue does the question of the practical effect of a ban become relevant. Given the concern shown by PMCs and PSCs over maintaining a good reputation, the majority of companies working in this area profess a very clear concern to abide by national law. The few companies that failed to do so would, of course, become objects of legal action.

33 C. Beyani and D. Lilly, op. cit.

34 Foreign and Commonwealth Office, op. cit., p. 23.

35 In interviews with the following persons, all supported the introduction of some type of licensing regime. Michael Grunberg, Official Spokesperson, Sandline International, London, 24 June 2002; Nic van den Berg, Operations Manager, Sandline International, Birmingham, 24 June 2002; Nigel Woof, Vice President Marketing, ArmorGroup, London, 30 November 2001. The human rights organisation International Alert also supports the introduction of a licensing regime see, Beyani and Lilly, op. cit.

36 Foreign and Commonwealth Office, op. cit., p. 24.

37 The Office of Defense Trade Controls (DTC), Bureau of Political–Military Affairs, in accordance with 38–40 of the Arms Export Control Act (AECA) (22 USC 27780) and the ITAR (22 CFR Parts 120–30), is charged with controlling the export and temporary import of defence articles and services covered by the US Munitions List.

38 Foreign Affairs Committee, op. cit., p. 33.

39 Christopher Beese, personal communication, 6 January 2004.

40 *United Kingdom Strategic Export Controls Annual Report 2003*.

41 Foreign Affairs Committee, op. cit., p. 33.

42 Christopher Beese, personal communication, 6 January 2004.

43 Ibid.

44 Interview with Michael Grunberg, London, 24 June 2002.

45 Foreign Affairs Committee, op. cit., Ev. 68.

46 Ibid., p. 25.

47 Christopher Beese, personal communication, 15 December 2003.

48 T. Catán and S. Fidler, 'The Military can't Provide Security. It had to be Outsourced to the Private Sector and that was our Opportunity', in *The Financial Times*, 29 September 2003.

49 The role of the private sector and human rights. Online. Available at http://www.business-humanrights.org/Categories/Principles/VoluntaryPrinciplesonSecurityHumanRights (accessed 1 June 2004).

50 Christopher Beese, personal communication, 15 December 2003.

51 For details of the 'Foreign Corrupt Practices Act' see http://www.usdoj.gov/criminal/fraud/fcpa.html (accessed 15 December 2003).

52 Christopher Beese, personal communication, 15 December 2003.

53 A number of companies already subscribe to such a code. In the case of ArmorGroup, for example, the company subscribes to the code of conduct of the International Red Cross and Red Crescent movement. Interview with Nigel Woof, Vice President Marketing, ArmorGroup, London, 30 November 2001.

54 UN document 'Basic Principles on the Use of Force and Firearms by Law Enforcement Officials', United States Department of State, 19 December 2000.

55 Voluntary Principles on Security and Human Rights Statement by the Governments of the United States of America and the United Kingdom, 19 December 2000.

56 Foreign and Commonwealth Office, op. cit., p. 26.

57 Ibid., p. 26.

58 P. Singer, op. cit., p. 548.

CONCLUSION

1 P. Singer, *Corporate Warriors: The Rise of the Privatized Military Industry*, London: Cornell University Press, 2003, pp. 88–100.

2 Radio Interview with Dr Shameem. Online. Available at http://www.goasiapacific.com/location/pacific/GAPLocPacificStories_1170422.htm (accessed 19 January 2005).

3 Jim Hooper spent 20 years studying the conventional–special forces combat operations of the SADF. He has written about these operations as well as those carried out by units of the Koevoet, the Southwest Africa Police Counterinsurgency unit, which was drawn primarily from the South African Police Security Branch, and which later made up the bulk of EO's mech-infantry troops. See J. Hooper, *Beneath the Visiting Moon: Images of Combat in Southern Africa*, America: Lexington Books, 1990.

4 Forças Armados de Popular de Libertaçac de Angola and Forças Armadas de Angola.

5 Interview with Ed Soyster, 10 September 2002.

6 The Polish Press Agency. Online. Available at http://www.kiosk.0net.pl/0,0784299,5470,przeglad.html (accessed on 27 August 2003).

7 K. Sengupta, 'Anglo–American security firm investigated over kidnap plot', in *Independent*, 8 August 2003.

8 C. Beyani and D. Lilly, *Regulating Private Military Companies: Options for the UK Government*, London: International Alert, 2001.

9 The Convention entered into force on 20 October 2001 in accordance with Article 19 (1) following the deposit of the 22 Instrument of Ratification on 20 September 2001.

10 The statutory authority of the President to promulgate regulations with respect to export of defence articles and defence services is delegated to the Secretary of State.

11 J. Thomson, *Mercenaries, Pirates, and Sovereigns*, Princeton, NJ: Princeton University Press, 1994, p. 146.

SELECTED BIBLIOGRAPHY

Books

Abrahamsen, R., *Disciplining Democracy*, London: Zed Books, 2000.

Alvesson, M., *The Management of Knowledge Intensive Companies*, Berlin: Walter de Gruyter, 1995.

Andrew, C., *For the President's Eyes Only*, London: Harper Collins, 1995.

Ayoob, M., *The Third World Security Predicament: State-Making, Regional Conflict, and the International System*, Boulder, CO: Lynne Rienner Publishers, 1995.

Baylis, J. and S. Smith (eds), *The Globalisation of World Politics*, Oxford: Oxford University Press, 1997.

Beevor, A., *Stalingrad*, London: Penguin, 1999.

Berdal, M. and D. Malone (eds), *Greed and Grievance: Economic Agendas in Civil Wars*, London: Lynne Rienner, 2000.

Black, J., *War for America*, Britain: Alan Sutton Publishing Ltd, 1991.

Bloch, J. and P. Fitzgerald, *British Intelligence and Covert Action*, London: Junction Books Ltd, 1983.

Boëne, B. and M.L. Martin, 'France: In the Throes of Epoch-Making Change' in, C. Moskos *et al.*, *The Post Modern Military: Armed Forces after the Cold War*, Oxford: Oxford University Press, 2000.

Boli, J. and G.M. Thomas (eds), *Constructing World Culture*, California: Stanford University Press, 1999.

Bower, T., *The Perfect English Spy*, London: Heinemann, 1995.

Boxer, C., *The Dutch Seaborne Empire 1600—1800*, London: Hutchinson & Co, 1972.

Brinton, C., in M. Earle *et al.*, *Makers of Modern Strategy*, Princeton, NJ: Princeton University Press, 1941.

Brown, R.A., *The Origins of Modern Europe*, London: Constable, 1972.

Buzan, B., *People States and Fear*, London: Harvester Wheatsheaf, 1991.

Challener, R., *The French Theory of the Nation in Arms*, New York, NY: Russell & Russell, 1965.

Clark, I., *Globalisation and Fragmentation: International Relations in the Twentieth Century*, Oxford: Oxford University Press, 1997.

Clark, W., *Waging Modern War*, Oxford: Public Affairs Ltd, 2001.

Cleaver, G., 'Subcontracting Military Power: The Privatisation of Security in Contemporary Sub-Saharan Africa', in *Crime, Law & Social Change*, Netherlands: Kluwer Academic Publishers, Vol. 33, 2000.

Coker, C., *Human Warfare*, London: Routledge, 2001.

Collier, P., *Economic Causes of Civil Conflict and their Implications for Policy*, Washington, DC: World Bank, 2000.

Coornaert, E., 'European Economic Institutions and the New World: The Chartered Companies', in E. Rich and C. Wilson (eds), *The Cambridge Economic History of Europe*, Cambridge: Cambridge University Press, 1967.

Crowhurst, P., *The Defence of British Trade, 1689—1815*, Folkestone: Dawson, 1977.

Curtiss, J., *Russia's Crimean War*, Durham, NC: Duke University Press, 1979.

Dandeker, C., 'The United Kingdom: The Overstretched Military', in C. Moskos *et al.*, *The Post Modern Military: Armed Forces after the Cold War*, Oxford: Oxford University Press, 2000.

Dandeker, C. and F. Paton, *The Military and Social Change: A Personnel Strategy for the British Armed Forces*, London: Brassey's, 1997.

Davies, H. and D. Holdcroft, *Jurisprudence: Text and Commentary*, London: Butterworths, 1991.

De Waal, A., *Famine Crimes: Politics and the Disaster Relief Industry in Africa*, London: African Rights & the International African Institute Oxford, in association with James Currey Bloomington: Indiana University Press, 1997.

Deacon, R., *British Secret Service*, London: Collins Publishing Group, 1991.

DiMaggio, P. and W. Powell, 'The Iron Cage Revisited: Institutional Isomorphism and Collective Rationality in Organisational Fields', in W. Powell and P. Dimaggio (eds), *The New Institutionalism in Organisational Analysis*, Chicago, IL: University of Chicago Press, 1991.

Dixon, N., *On the Psychology of Military Incompetence*, London: Jonathan Cape, 1976.

Dorril, S., *MI6: Fifty Years of Special Operations*, London: Fourth Estate, 2000.

Douglas, I., 'Fighting For Diamonds: Private Military Companies in Sierra Leone', in J. Cilliers and P. Mason (eds), *Peace, Profit, or Plunder*, South Africa: Institute for Security Studies, 1999.

Duffield, M., *Global Governance and the New Wars*, London: Zed Books, 2001.

Dunne, T., 'Realism', in J. Baylis and S. Smith (eds), *The Globalisation of World Politics*, Oxford: Oxford University Press, 1997.

Farer, T. and F. Gaer, 'The UN and Human Rights: At the End of the Beginning', in A. Roberts and B. Kingsbury (eds), *United Nations, Divided World*, Oxford: Clarendon Press, 1995.

Finnemore, M., 'Rules of War and Wars of Rules: The International Red Cross, and the Restraint of State Violence', in J. Boli and G. Thomas, *Constructing World Culture*, California, CA: Stanford University Press, 1999.

Furber, H., 'Rival Empires of Trade in the Orient, 1600—1800', in B. Shafter, *Europe and the World in the Age of Expansion*, Oxford: Oxford University Press, 1976.

Gates, B., *Warfare in the Nineteenth Century*, Basingstoke: Palgrave, 2001.

Gibbins, J.R., 'Contemporary Political Culture: an Introduction' in, J.R. Gibbins (ed.), *Contemporary Political Culture: Politics in a Postmodern Age*, London: Sage Publications, Vol. 23, 1990.

Giddens, A., *A Contemporary Critique of Historical Materialism, Volume 2, The Nation State and Violence*, California, CA: University of California Press, 1985.

Gilpin, R., *The Challenge of Global Capitalism: The World Economy in the 21 Century*, Princeton, NJ: Princeton University Press, 2000.

Gordon, J., *Organisational Behaviour: A Diagnostic Approach*, London: Prentice-Hall International, 1999.

Hackett, J., 'Society and the Soldier', in M. Wakin (ed.), *War, Morality and the Military Profession*, London: Westview Press, 1986.

Hall, W.E., *A Treatise on International Law*, Oxford: Clarendon Press, 1924.

Hampson, F.J., 'Mercenaries: Diagnosis before Proscription', in Netherlands Yearbook of International Law, Vol. 22, No. 3, Netherlands: Martinus Nijhoff Publishers, 1991.

Haufler, V., *A Public Role for the Private Sector*, Washington, DC: Carnegie Endowment for International Peace, 2001.

Hirsch, J., *Sierra Leone: Diamonds and the Struggle for Democracy*, London: Lynne Rienner, 2001.

Hirst, P. and G. Thompson, *Globalisation in Question*, Cambridge: Polity Press, 1999.

Hobsbawm E.,, *Age of Extremes*, London: Abacus, 1997.

Hoe, A., *David Stirling: The Authorised Biography of the Creator of the SAS*, London: Warner Books, 1992.

Hooper, J., *Bloodsong!*, London: Harper Collins Publishers, 2002.

Hopkins, T. and I. Wallerstein (eds), *The Age of Transition*, London: Zed Books, 1996.

Howard, M., *War in European History*, Oxford: Oxford University Press, 1977.

Howe, H., *Ambiguous Order: Military Forces in African States*, London: Lynne Rienner Publishers, 2001.

Human Rights Watch, *Angola: Arms Trade and the Violations of the Laws of War*, New York, NY: Human Rights Watch, 1994.

Hutchful, E., 'Understanding the African Security Crisis', in A-F Musah and J. Fayemi, *Mercenaries*, London: Pluto Press, 2000.

Ignatieff, M., *The Warrior's Honor: Ethnic War and the Modern Conscience*, London: Chatto & Windus, 1998.

Jackson, R., *Quasi-States: Sovereignty, International Relations, and the Third World*, New York, NY: Cambridge University Press, 1990.

Janowitz, M. and R. Little, *Sociology and the Military Establishment*, London: Sage Publications, 1974.

Jaques, E., *The Changing Culture of a Factory*, London: Tavistock, 1952.

Jepperson, R., 'Institutions, Institutional Effects, and Institutionalism', in W. Powell and P. DiMaggio (eds), *The New Institutionalism In Organisational Analysis*, Chicago, IL: University of Chicago Press, 1991.

Jolly, R., *Military Man, Family Man: Crown Property*, London: Brassey's Defence Publishers, 1987.

Kaldor, M., 'Introduction', in M. Kaldor and B. Vashee (eds), *New Wars*, London: Pinter, 1997.

Kaldor, M., *New & Old Wars*, Cambridge: Polity Press, 2002.

Keegan, J., *A History of Warfare*, London: Hutchinson, 1993.

Keegan, J., *War and Our World*, London: Pimlico, 1999.

Keen, D., *The Economic Function of Violence in Civil Wars*, Adelphi Paper 320, London: International Institute for Strategic Studies, 1998.

Keen, D., 'Incentives and Disincentives for Violence', in M. Berdal and D. Malone, *Greed and Grievance: Economic Agendas in Civil Wars*, London: Lynne Rienner Publishers, 2000.

Kegley, C. and E. Wittkopf, *World Politics Trends and Transformation*, London: The Macmillan Press Ltd, 1995.

Kennedy, P., *The Rise and Fall of the Great Powers*, London: Fontana Press, 1988.

Keohane, O.R. and J. Nye (eds), *Transnational Relations and World Politics*, Cambridge, MA: Harvard University Press, 1981.

Kissinger, H., *Years of Renewal*, London: Weidenfeld & Nicholson, 1999.

Krasner, S., *Sovereignty Organised Hypocrisy*, Princeton, NJ: Princeton University Press, 1999.

Le Caron, H., *Twenty Five Years in the Secret Service: The Recollections of a Spy*, Wakefield: EP Publishing, 1974.

Leslie, S., *The Cold War and American Science*, New York, NY: Columbia University Press, 1993.

McAleese, P., *No Mean Soldier*, London: Orion, 1998.

MacIver, R., *The Modern State*, Oxford: Oxford University Press, 1964.

Mackay, D., *The Honourable Company: A History of the Hudson Bay Company*, New York, NJ: Bobbs Merill, 1936.

McNeill, W., *The Pursuit of Power*, Basil: Blackwell Publishers Ltd, 1984.

Mandel, R., *Armies without States: the Privatisation of Security*, London: Lynne Rienner Publishers, 2002.

Mann, M., *The Source of Social Power*, Cambridge: Cambridge University Press, Volume 2, 1993.

Martin, J., *Organisational Behaviour*, London: International Thomson Business Press, 1998.

Martin, L. and J. Garnett, *British Foreign Policy: Challenges and Choices for the 21st Century*, London: The Royal Institute of International Affairs, 1997.

May, T., *Social Research: Issues, Methods and Process*, London: Open University Press, 2001.

Melvern, L., *A People Betrayed*, London: Zed Books, 2000.

Meyer, J. and B. Rowan, 'Institutionalised Organisations: Formal Structure as Myth and Ceremony', in W. Powell and P. Dimaggio (eds), *The New Institutionalism In Organisational Analysis*, Chicago, IL: University of Chicago, 1991.

Mockler, A., *The New Mercenaries*, New York, NY: Macmillan, 1986.

Moller, B., 'Conscription and its Alternatives', in *The Comparative Study of Conscription in the Armed Forces*, The Netherlands: Elsevier Science Ltd, Vol. 20, 2002.

Moskos, C., 'Towards a Postmodern Military: The United States', in C. Moskos *et al.*, *The Post Modern Military: Armed Forces after the Cold War*, Oxford: Oxford University Press, 2000.

Musah, A.-F. and J. Fayemi (eds), *Mercenaries*, London: Pluto Press, 2000.

O'Brien, K., 'Private Military Companies and African Security' in A-F. Musah and J. Fayemi, *Mercenaries*, London: Pluto Press, 2000.

Olson, M., *The Logic of Collective Action*, Cambridge, MA: Harvard University Press, 1965.

Paret, P., 'Clausewitz', in P. Paret (ed.), *Makers of Modern Strategy: From Machiavelli to the Nuclear Age*, Oxford: Clarendon Press, 1991.

Paxman, J., *Friends in High Places*, London: Penguin Books, 1991.

Pech, K., 'Executive Outcomes: A Corporate Conquest', in J. Cilliers and P. Mason, *Peace, Profit or Plunder*, South Africa: Institute for Security Studies, 1999.

Powell, W., 'Expanding the Scope of Institutional Analysis', in W. Powell and P. DiMaggio (eds), *The New Institutionalism in Organisational Analysis*, Chicago, IL: University of Chicago Press, 1991.

Rapoport, A., 'Introduction to On War,' in C. Von Clausewitz, *On War*, London: Penguin Classics, 1982.

Reno, W., *Warlord Politics and African States*, London: Lynne Rienner, 1998.

184

Reno, W., 'Shadow States and the Political Economy of Civil Wars', in M. Berdal and D. Malone, *Greed and Grievance: Economic Agendas in Civil Wars*, London: Lynne Rienner, 2000.

Roberts, A., 'The Crisis in UN Peacekeeping', in C. Crocker and F. Hampson, *Managing Global Chaos*, Washington, DC: Institute of Peace, 1996.

Roberts, A. and R. Guelff, *Documents on the Laws of War*, Oxford: Oxford University Press, 2000.

Roberts, A. and B. Kingsbury (eds), *United Nations, Divided World*, Oxford: Clarendon Press, 1995.

Roberts, M., 'The Military Revolution, 1560—1660', in C. Rogers (eds), *The Military Revolution Debate*, Oxford: Westview Press Inc, 1995.

Sanders, D., *Losing an Empire, Finding a Role*, London: Macmillan Press Ltd, 1990.

Sarkesian, S. and R. Connor, *The US Military Profession into the Twenty First Century*, London: Frank Cass, 1999.

Scott, R., 'Unpacking Institutional Arguments', in W. Powell and P. DiMaggio, *The New Institutionalism in Organisational Analysis*, Chicago, IL: University of Chicago Press, 1991.

Segal, M., 'The Argument for Female Combatants', in N. Goldman, *Female Soldiers: Combatants or Non-Combatants?*, London: Greenwood, 1982.

Shaw, M., *Post Military Society*, Cambridge, Polity Press, 1991.

Shaw, M.N., *International Law*, Cambridge: Cambridge University Press, 1995.

Shearer, D., 'Private Armies and Military Intervention', in *Adelphi Paper 316*, London: International Institute for Strategic Studies, 1998.

Sherry, F., *Raiders and Rebels: The Golden Age of Piracy*, New York, NY: Hearst Marine Books, 1986.

Silverstein, K., *Private Warriors*, London: Verso, 2000.

Singer, P., *Corporate Warriors: The Rise of the Privatised Military Industry*, London: Cornell University Press, 2003.

Spicer, T., *An Unorthodox Soldier: Peace and War and the Sandline Affair*, Edinburgh: Mainstream Publishing Company, 1999.

Stark, F., *The Abolition of Privateering and the Declaration of Paris*, New York, NY: Columbia University, 1897.

Symcox, G., *The Crisis of French Sea Power 1688—97*, The Hague: Martinus Nijhoff, 1974.

Thomson, J., *Mercenaries, Pirates, and Sovereigns*, Princeton, NJ: Princeton University Press, 1994.

Thurlow, R., *The Secret State*, Oxford: Blackwell Publishers, 1994.

Tilly, C., *The Formation of Nation States in Western Europe*, Princeton, NJ: Princeton University Press, 1975.

Tilly, C., *Coercion, Capital, and European States, AD 990—1992*, Oxford: Blackwell Publishers, 1992.

Van Creveld, M., *Command In War*, Cambridge, MA: Harvard University Press, 1985.

Van Creveld, M., *Supplying War*, Cambridge: Cambridge University Press, 1992.

Van Creveld, M., 'The Transformation of War Revisited', in R.J. Bunker (ed.), *Non-State Threats and Future Wars*, London: Frank Cass, 2003.

Vines A., 'Gurkhas and the Private Security Business in Africa', in J. Cilliers and P. Mason, *Peace, Profit, or Plunder*, South Africa: Institute for Security Studies, 1999.

Vines, A., 'Mercenaries and the Privatisation of Security in Africa in the 1990s', in *The Privatisation of Security in Africa*, South Africa: The South African Institute of International Affairs, 1999.

Vines, A., 'Mercenaries, Human Rights and Legality', in A-F. Musah and J. Fayemi, *Mercenaries*, London: Pluto Press, 2000.

Wavell, A., *Soldiers and Soldiering*, London, 1953.

Weber, M., *The Theory of Social and Economic Organisation*, translation by A.M. Henderson and Talcott Parsons, New York, NY: Free Press, 1964.

Weber, M., *Economy and Society*, Berkeley, CA: University of California Press, 1978.

Wheeler, N., *Saving Strangers: Humanitarian Intervention in International Society*, Oxford: Oxford University Press, 2000.

Williams, J., 'The Postmodern Military Reconsidered', in C. Moskos *et al.*, *The Post Modern Military: Armed Forces after the Cold War*, Oxford: Oxford University Press, 2000.

Journals

Brooks, D., 'Managing Armed Conflict in the 21 Century', in *International Peacekeeping*, Vol. 7, No. 4, 2000.

Callaghy,, T. 'The State–Society Struggle: Zaire', in *Comparative Perspective*, New York, NY: Columbia University Press, 1984.

Ficarrotta, J., 'Are Military Professionals Bound by a Higher Moral Standard', in *Armed Forces Society*, US: Transaction Periodicals Consortium – Rutgers University, Vol. 24, No. 1, 1997.

Higate, P., 'Theorizing Continuity: From Military to Civilian Life', in *Armed Forces Society*, US: Transaction Periodicals Consortium – Rutgers University, Vol. 27, No. 3, 2001.

Howe, H., 'Private Security Force and African Stability: The Case of Executive Outcomes', in *The Journal of Modern African Studies*, Cambridge: Cambridge University Press, Vol. 36 No. 2, 1998.

Hurrell, A. 'Explaining the Resurgence of Regionalism in World Politics', in *Review of International Studies*, London: Butterworths, Vol. 21, No. 4, 1995.

Kinsey, C., 'Private Military Companies: Options for Regulation', in *Conflict, Security, Development*, London: King's College London, Vol. 2, No. 3, 2002.

McGhie,, S. 'Private Military Companies: Soldiers, Inc', in *Jane's Weekly Defence*, London: Thanet Press Ltd, Vol. 37, No. 21, 2002.

Meyer, J., 'Institutionalised Organisations: Formal Structure as Myth and Ceremony', in *American Journal of Sociology*, Chicago, IL: University of Chicago Press, 1977.

Meyer, M., and M. Brown, 'The Process of Bureaucratisation', in *American Journal of Sociology*, Chicago, IL: University of Chicago Press, 1977.

O'Brien, K., 'Military–Advisory Groups and African Security: Privatised Peacekeeping', in *International Peacekeeping*, London: Frank Cass Vol. 5, 1998.

Reyntjens, F., 'The Second Congo War: More Than A Remake', in *African Affairs*, Oxford: Oxford University Press, Vol. 98, No. 391, 1999.

Shearer, D., 'Outsourcing War', in *Foreign Policy*, Washington DC: Carnegie Endowment for International Peace, 1998.

Shearer, D., 'Private Military Companies and Challenges for the Future', in *Cambridge Review of International Relations*, Vol. XIII, No. 1, 1999.

Singer,, P. 'Corporate Warriors: The Rise of the Private Military Industry and Its Ramifications for International Security', in *International Security*, MA: Harvard University, Vol. 26, No. 3, 2002.

Smith, C., 'Security Sector Reform: Development Breakthrough or Institutional Engineering?', in *Conflict Security Development*, London: King's College London, 2001.

Spearin, C., 'Private Security Companies and Humanitarianism: A Corporate Solution to Security Humanitarian Space', in *International Peacekeeping*, Vol. 8, No. 1, 2001.

Spicer, T., 'Interview', in *Cambridge Review of International Affairs*, Cambridge: Vol. XIII, No. 1, 1999.

Government documents

Diplock, Lord, Sir Derek Walker Smith and Sir Geoffrey de Freitas, *The Diplock Report*, Command 6569 (XXI), London: The Stationery Office, 1976.

Foreign Affairs Committee, 'Second Report', *Sierra Leone*, London: The Stationery Office, 1999.

Foreign and Commonwealth Office, 'Private Military Companies: Options for Regulation', *The Green Paper*, London: The Stationery Office, 2002.

House of Commons Foreign Affairs Committee, *Private Military Companies*, London: The Stationery Office, 23 July 2002.

Legg, Sir Thomas, KCB QC and Sir Robin Ibbs KBC, *Report Of The Sierra Leone Arms Investigation*, London: The Stationery Office, 1998.

United Nations documents

Ballesteros, E., *Report on the question of the use of mercenaries as a means of violating human rights and impeding the exercise of the right of peoples to self-determination*, Commission Resolution, 1998/6.E/CN.4/1999/11, Geneva, The Office of the High Commissioner for Human Rights, 1999.

Ballesteros, E., *Use of Mercenaries as Means of Violating Human Rights and Impeding the Exercise of the Right of People to Self Determination*, A/54/326, Geneva, The Office of the High Commissioner for Human Rights, 1999.

United Nations Security Council No. S/1997/499 (1997).

United Nations Security Council Resolution No. 1132 (1997).

General documents

Beyani, C. and D. Lilly, *Regulating Private Military Companies: Options for the UK Government*, London: International Alert, 2001.

International Alert, *Private Military Companies and the Proliferation of Small Arms: Regulating the Actors*, London: International Alert, 2000.

Isenberg, D., 'A Profile of Today's Private Sector Corporate Mercenary Firms', in Center for Defense Information Monograph, Washington, DC: Center for Defense Information, 1997.

Lilly, D., *Regulating Private Military Companies: The Need for a Multidimensional Approach*, London: International Alert, 2002.

Vaux, T., *et al.*, *Humanitarian Action and Private Security Companies*, London: International Alert, 2002.

Newspapers

Bazargan, D., 'High-risk business', in *The Guardian*, 8 September 1997.

Bilton, M., 'Death and Dollars', in *The Sunday Times Magazine*, 2 July 2000.

Bilton, M., 'The Private War of Tumbledown Tim', in *Sunday Times Magazine*, 2 July 2000.

Black, I., 'International Criminal Court Comes to Life', in *The Guardian*, 11 March 2003.

Bonner, R., 'French Covert Action in Zaire on behalf of Mbutu', in *New York Times*, 2 May 1997.

Burns, J. and A. Parker, 'Ex-SAS soldiers hope for new lease of life', in *The Financial Times*, 14 February 2002.

Catán, T. and S. Fidler, 'The Military Can't Provide Security. It had to be Outsourced to the Private Sector and that was our Opportunity', in *The Financial Times*, 29 September 2003.

Catán, T. and M. Peel, 'FBI Investigates Private Offer to Arrest Liberian President', in *The Financial Times*, 6 August 2003.

Colvin, M. and N. Rufford, 'Our Man in Sierra Leone is the People's Hero', in *The Times*, 17 May 1998.

Dyer, C., 'Compensation for Gays Sacked from Forces Stymied', in *The Guardian*, 24 October 2002.

Editorial, 'DynCorp's Assignment: Protect Afgahan Leader', in *The Washington Post*, 2 December 2002.

Editorial, 'On the Backs of Bosnians', in *The Washington Post*, 2 July 2002.

Editorial, 'Peacekeepers used Women as Sex Slaves', in *The Daily Mail*, 24 April 2002.

Inside Story, 'BP's Secret Soldiers', in *The Guardian*, 4 July 1997.

Isenberg, D., 'There's no Business like Security Business', in *Asia Times*, 30 April 2003.

Isenberg D., 'Security for Sales', in *Asia Times*, 14 August 2003.

Kagan, R., 'Europeans Courting International Disaster', in *The Washington Post*, 30 June 2002.

Kurlantzick, 'Outsourcing the Dirty Work; The Military and its Reliance on Hired Guns', in *The New Republic*, 24 April 2003.

La Guardia, A., 'Mercenaries May be Key to Oust Taylor', in *The Daily Telegraph*, 25 July 2003.

Lee, C. 'Army Weighs Privatisation Close to 214,000 Jobs One in Six Workers could be Affected', in *The Washington Post*, 3 November 2002.

Lee, C., 'Army Outsourcing Plan Decried Thousands of Jobs could be Affected', in *The Washington Post*, 21 December 2002.

Little, R., 'American Civilians Go Off To War', in *Baltimore Sun*, 26 May 2002.

Marx, G., 'U.S. Civilians Wage Drug War from Columbia's Skies Program Strives to Eradicate Coca', in *Chicago Tribune*, 3 November 2002.

Mathiason, N., 'Private Contractors are Carving up Defence Procurement', in *The Observer*, 2 March 2003.

Meadow, J., 'This War Game May be a Lifesaver; Soldiers Learn the Ropes out of Harm's Ways in Simulator', in *Rocky Mountain News*, 25 January 2003.

Millis, W., 'Puzzle of the Military Mind', in *New York Times*, 18 November 1972.

Morgan, O., 'Soldiers of Fortune Hit the Jackpot', in *The Observer*, 26 October 2003.

O'Connor, P., 'Australia Probes East Timor Accusation', in *The Associated Press*, 3 October 2002.

O'Loughlin, E., 'Sandline Scandal Arms Shipment Reaches Forces', in *The Independent*, 22 May 2000.

O'Meara, K., 'The Last Word: DynCorp Still Taking the Moral Low Ground', in *Insight Magazine*, 2 September 2002.

Payne, S., 'Teenagers Used for Sex By UN in Bosnia', in *The Daily Telegraph*, 25 April 2002.

Sengupta, K., 'Anglo–American Security Firm Investigated Over Kidnap Plot', in *The Independent*, 8 August 2003.

Tamayo, J., 'U.S. Civilians on Frontlines of Drug War in Colombia', in *Knight Rider Newspaper*, 26 February 2001.

Tamayo, J., 'Pentagon Contractors Playing Vital Military Support Role', in *Miami Herald*, 7 March 2003.

Tepperman, J., 'Out of Service', in *New Republic*, 13 January 2003.

Thornycroft, P., 'Mobutu Couldn't Afford SA Mercenaries', in *The Electronic Mail & Guardian*, 18 July 1997.

Washington in Brief, 'Use of Private Guards for Karzai Criticized', in *The Washington Post*, 18 September 2002.

Waugh, P. and N. Morris, 'Mercenaries as Peacekeepers', in *The Independent*, 14 February 2002.

INDEX